THE ZEN OF
MAGIC SQUARES,
CIRCLES, AND STARS

Also by Clifford A. Pickover

The Alien IQ Test
Black Holes: A Traveler's Guide
Chaos and Fractals
Chaos in Wonderland
Computers and the Imagination
Computers, Pattern, Chaos, and Beauty
Cryptorunes
Dreaming the Future
Fractal Horizons: The Future Use of Fractals
Frontiers of Scientific Visualization
    (with Stuart Tewksbury)
Future Health: Computers and Medicine
    in the 21st Century
The Girl Who Gave Birth to Rabbits
Keys to Infinity
The Loom of God
Mazes for the Mind: Computers and the Unexpected
The Paradox of God and the Science of Omniscience
The Pattern Book: Fractals, Art, and Nature
The Science of Aliens
Spider Legs (with Piers Anthony)
Spiral Symmetry (with Istvan Hargittai)
The Stars of Heaven
Strange Brains and Genius
Surfing Through Hyperspace
Time: A Traveler's Guide
Visions of the Future
Visualizing Biological Information
Wonders of Numbers

THE ZEN OF MAGIC SQUARES, CIRCLES, AND STARS

An Exhibition of Surprising Structures across Dimensions

Clifford A. Pickover

Princeton University Press

Princeton and Oxford

Published by Princeton University Press, 41 William Street,
Princeton, New Jersey 08540

In the United Kingdom: Princeton University Press, 3 Market Place, Woodstock,
Oxfordshire OX20 1SY

Library of Congress Cataloging-in-Publication Data

Pickover, Clifford A.
The zen of magic squares, circles, and stars : an exhibition of surprising structures
across dimensions / Clifford A. Pickover.
   p. cm
Includes bibliographical references and index.
ISBN 0-691-07041-5 (acid-free paper)
1. Magic squares. 2. Mathematical recreations.   I. Title.
QA165.P53   2002
511'.64–dc21   2001027848

British Library Cataloging-in-Publication Data is available

This book has been composed in Baskerville BE and Gill Sans.

Printed on acid-free paper ∞

www.pup.princeton.edu

Printed in the United States of America
10 9 8 7 6 5 4 3 2 1

The peculiar interest of magic squares lies in the fact
that they possess the charm of mystery.
They appear to betray some hidden intelligence
which by a preconceived plan produces
the impression of intentional design,
a phenomenon which finds its
close analogue in nature.

—Paul Carus, in W. S. Andrews's *Magic Squares and Cubes*

The mathematical phenomenon always develops
out of simple arithmetic, so useful in everyday life,
out of numbers, those weapons of the gods:
the gods are there, behind the wall,
at play with numbers.

—Le Corbusier, *The Modulor*

To study magic squares is to study the self.
To study the self is to forget the self.
To forget the self is to be enlightened.

—Abhinavagupta Isvarapratyabhijna[1]

The magic square is the hammer that
shatters the ice of our unconscious.

—Qingfu Chuzhen[2]

Translation: the Zen of magic squares. The phrase "magic squares" is written literally as "square puzzles." On its own, this Chinese word for puzzle refers to the location of soldiers and weapons on a battlefield as described in Sun Zi's Art of War. However, together with the Chinese word for "square," the phrase denotes "magic squares." The small square at the bottom encloses the name of the calligrapher, Siu-Leung Lee.

This book is dedicated not to a person
but to a meditative aid,
the Durga Yantra,
to which numbers
can be applied
in magic
ways.

Contents

Preface

Art and science will eventually be seen to be as closely connected as arms to the body. Both are vital elements of order and its discovery. The word "art" derives from the Indo-European base "ar," meaning to join or fit together. In this sense, science, in the attempt to learn how and why things fit, becomes art. And when art is seen as the ability to do, make, apply or portray in a way that withstands the test of time, its connection with science becomes more clear.

–Sven Carlson, *Science News*

If we wish to understand the nature of the Universe we have an inner hidden advantage: we are ourselves little portions of the universe and so carry the answer within us.

–Jacques Boivin, *The Single Heart Field Theory*

Lepidoptera Overdrive

Sometimes you are a butterfly collector, a person with a net. You don't always know what the meadows will yield, but you know when the breeze is right, where the meadows are fertile, and what type of net to use. Often the specific catch is a surprise, and this is the enjoyment of the quest. There are no guarantees. There are often unexpected pleasures.

Follow me as we search through unknown forests. Each structure we uncover contains magnificent patterns as beautiful as any swallowtail's wing. Hopefully you will enjoy looking at the catches or magnifying them further to learn more about their internal structures. Sometimes you may feel as if you are walking through an exhibit of butterflies in a gallery. At other times you may feel as if you are standing on the edge of a gigantic crystal, surrounded by thousands of butterfly wings beating in unison, producing a hum that reminds you of the chanting of monks.

Beyond Magic Squares

Magic squares have fascinated humans since the dawn of civilization. Even in ancient Babylonian times, people considered these squares to have magical powers, and in the eighth century A.D., some squares were considered useful for turning ordinary metal into gold. The patterns have also been used as religious symbols, protective charms, and tools for divination. When the squares lost their mystical meanings, laypeople continued to use them as fascinating puzzles, while seasoned mathematicians studied them as problems in number theory. Albrecht Dürer, the fourteenth-century painter and printmaker, used them in his artworks, and today magic squares continue to intrigue us with their elegant, beautiful, and strange symmetries.

A *magic square* is a square array of integers in which the rows, columns, and diagonals have the same sums. However, in this book we will go far beyond ordinary magic squares and consider many unusual variations, some in higher dimensions, all with mind-boggling patterns. Many of the squares possess an intrinsic beauty and complexity hard to describe or understand without careful study. For example, some squares remind me of fractals, geometric patterns with similar structures contained within larger structures. The numerical diagrams tantalize us with hidden patterns as if some mathematician-God played a role in their design. In this book, you will find squares within squares; cubes within cubes; four-dimensional objects; patterns based on ancient, sacred geometry; and structures that defy classification. Mathematician Benoit Mandlebrot likens the creation of nested structures to Zen philosophy:

> They are repeated without end. Infinite regression is something fully described in Zen Buddhist books, and often appears in the works of Leibniz and Kant.[1]

While thinking about these perplexing patterns, you may be at first confused, but then you will find that your mind "clicks" as you suddenly appreciate the structure. It is not an exaggeration to think of this moment of understanding as a miniature epiphany, a revelation, a piece of *satori* in which each magic structure is a mandala for the mind. Due to space limitations, I sometimes exhibit just a few hidden patterns to plant a seed, allowing you to consult the references or explore with a computer to make additional discoveries.

Revelation

In my mind, we don't invent magic squares, but rather we discover them. Magic squares are out there in the realm of eternal ideas. They have an independent existence from us. These ideas are controversial, and there are certainly other points of view.[2] However, to me, magic squares transcend us and our physical reality. The statement "A magic square has particular symmetries and properties" is either true or false. As you look through this book, you can see that it turns out to be true. Was the statement true before the invention and discovery of magic squares? I believe it was. Magic squares exist whether humans know about them or not.

I think mathematics is a process of discovery. Mathematicians are like archeologists. Physicist Roger Penrose felt the same way about fractal geometry. In his book *The Emperor's New Mind*, he says that fractals (for example, intricate patterns such as the Julia set or Mandelbrot set) are out there waiting to be found:

> It would seem that the Mandelbrot set is not just part of our minds, but it has a reality of its own ... The computer is being used in essentially the same way that an experimental physicist uses a piece of experimental apparatus to explore the structure of the physical world. The Mandelbrot set is not an invention of the human mind: it was a discovery. Like Mount Everest, the Mandelbrot set is just there.[3]

I think we are uncovering truths and ideas independent of the computer or mathematical tools we have invented. Penrose went a step further about fractals:

> When one sees a mathematical truth, one's consciousness breaks through into this world of ideas. . . . One may take the view that in such cases the mathematicians have stumbled upon works of God.[4]

I wonder whether magic squares are not touching the very fabric of our brains. Could it be that a magic square is of such extraordinary richness that it is bound to resonate with our neuronal networks and elicit pure pleasure? And if you agree that magic squares are marvelous, you should keep in mind that many other ideas in mathematics can elicit an even greater philosophical wonder.[5]

Zen Buddhists have developed questions and statements called *koans* that function as a meditative discipline. Koans ready the mind so that it can entertain new intuitions, perceptions, and ideas. Koans cannot be answered in ordinary ways because they are paradoxical; they function as tools for enlightenment because they jar the mind. Similarly, the contemplation of magic squares, and their bizarre geometrical cousins, is replete with koans, and that is why I am excited to present the following gallery of forms. These patterns are koans for scientific minds. I enjoy meditating on them.

Arithmetic satori is the psychological result and aim of the practice of magic square meditation. At the risk of appearing overly mystical, let me quietly say that this practice induces an awareness, an experience of joy emanating from a mind that has transcended its earthly existence. Experience is no longer mediated through concepts–which is why it is difficult to define arithmetic satori. In addition, the satori experience has a paradoxical quality, such as a feeling of oneness, that is inexpressible in a language posited on a subject-object dichotomy. The existence of a separate self is viewed as a fiction through the satori experience. Awareness seems to take place directly, unmediated by conscious thought, and without consciousness of the process.

Okay, some of you are thinking this sounds nuts, like something Timothy Leary[6] might have said while under the influence of LSD. But permit me just a few more thoughts on this subject, and I'll leave it forever. . . . Arithmetic satori may be experienced for shorter or longer times, depending on the length of training and the responsiveness of the individual to it. Brief experiences with a few exceptional magic squares in higher dimensions may be developed so that the experience occurs in a wider variety of conditions. Flashes of satori in arithmetic Zen are referred to as *numerical kensho*. Enlightenment, a nearly impossible ideal, is considered to be the constant experience of satori.

Smorgasbord

The mysterious, odd, and *fun* patterns in this book should cause even the most left-brained readers to fall in love with numbers. Some patterns are world records and are presented here for the first time. Other structures are centuries old. When I talk to students about the strange arrays in this book, they are always fascinated to learn that it is possible for them to break magic square world records and make new discoveries with a hand calculator. Most of the ideas can be explored with just a pencil and paper! You can sometimes discover new patterns in classic magic squares that were never known previously.

Even the famous eighteenth-century American Benjamin Franklin was fascinated by magic squares, although he once considered them a waste of time. Franklin wrote that "it was perhaps a mark of the good sense of our mathematicians that they would not spend their time in things that were merely *difficiles nuage*, incapable of any useful application."[7] But he then admitted to having carefully studied and composed some amazing magic squares, even going so far as to declare one square "the most magically magical of any magic square ever made by any magician."[8] You'll find this wonderful square in chapter 3.

One of the abiding sins of mathematicians is an obsession with completeness—an urge to go back to first principles to explain their works. As a result, readers must often wade through pages of background before getting to the essential ingredients. To avoid this problem, each "gallery exhibit" in chapters 3 through 5 is short. One advantage of this format is that you can jump right in to experiment with and enjoy the patterns. This book is not intended for mathematicians looking for formal mathematical explanations. Of course, this approach has some disadvantages. In just a few pages, I can't go into any depth on a subject. You won't find much historical context or extended discussions. However, I provide lots of extra material in the Notes and For Further Reading sections.

To some extent, the choice of topics for inclusion in this book is arbitrary. However, the topics chosen give a nice introduction to some common and unusual problems concerning magic squares. Many squares are representative of a wider class of structures of interest to mathematicians and puzzlists today. Some information is repeated so that gallery exhibits can be studied at random. The exhibits vary in difficulty, so you are free to browse.

Prepare yourself for a strange journey as *The Zen of Magic Squares, Circles, and Stars* unlocks the doors of your imagination. Some of the topics in this book may appear to be curiosities, with little practical application or purpose. However, I have found these experiments to be useful and educational—as have the many students, educators, and scientists who have written to me. Throughout history, experiments, ideas, and conclusions originating in the play of the mind have found striking and unexpected practical applications. A few possible practical applications are hinted at in several gallery exhibits.

This book will allow you to travel through time and space. To facilitate your journey, I have scoured the Earth in a quest for unusual people and their fascinating magic squares. From Benjamin Franklin's "most magical" magic square to John Hendricks's four-dimensional magic tesseracts, the exhibitors range from world-famous politicians and scientists, to little-known artists, to eclectic computer programmers.

To start you on the journey, I'll first provide you with some relevant background material on magic square creation, classification, and history and graphical representations that can be quite beautiful. I think you will find that the line between mathematics and art is a fuzzy one; the two are fraternal philosophies formalized by ancient Greeks such as Pythagoras and Ictinus. Today, magic square *geometrical diagrams* are methods by which mathematicians and artists reunite these philosophies by providing scientific ways to represent abstract objects.

In deciding how to arrange material within the chapters of *The Zen of Magic Squares, Circles, and Stars,* many divisions came to mind—linear and circular forms, computer- and non-computer-generated patterns, ancient and modern squares. However, the line between these categories becomes indistinct or artificial, and I have therefore randomly arranged the patterns within chapters to retain the playful spirit of this book and to give you unexpected pleasures. Some patterns could easily be placed in more than one chapter of this book. You are free to pick and choose from the smorgasbord of designs.

The literature on magic squares is vast, most of it written by laypeople who became addicted to the elegant symmetries of these nested number patterns. Several classic books on magic squares have been published, such as W. S. Andrews's *Magic Squares and Cubes* published in the early 1900s,[9] but these books obviously do not include some of the magnificent structures discovered in the past few years. I know of no book that presents such a large range of patterns as those included here. There are, however, several excellent books published on magic squares that are listed in the For Further Reading section. I think you will enjoy these.

Acknowledgments

She is light itself and transcendent. Emanating from her body are rays in thousands—two thousand, a hundred thousand, tens of millions, a hundred million—there is no counting their numbers. It is by and through Her that all things moving and motionless shine. It is by the light of this Devi that all things become manifest.

–Bhairava Yamala[1]

I owe a special debt of gratitude to the late W. S. Andrews and Paul Carus, editor and contributor to the book *Magic Squares and Cubes*, from which I have drawn many facts about these fascinating patterns. I also highly recommend John Lee Fults's *Magic Squares* for further reading. The various works of Martin Gardner and Joseph Madachy, listed in the For Further Reading section, have also been influential in my formulating an eclectic view of magic squares.

A number of the magic squares and related geometries exhibited in this book are from the remarkable John Hendricks, one of the world's foremost experts on magic squares. John started collecting magic squares and cubes when he was thirteen years old. His hobby became an obsession, and he soon became the first person to successfully make and publish four-, five-, and six-dimensional models of magic hypercubes. He is also the first person to exhibit inlaid magic cubes that consist of magic cubes within magic cubes.

The Chinese calligraphy before the dedication was contributed by Dr. Siu-Leung Lee. Dr. Lee has been practicing the art of calligraphy for more than forty years. Capable of writing in many styles, Dr. Lee has created his own style evolving from those of the Han and Jin dynasties. The calligraphy combines archaic structure and fluid movements to symbolize the dynamic nature of the universe. See his web site asiawind.com.

The Durga Yantra on the dedication page and in Figure 5.18 is by PennyLea Morris Seferovich. The mandalas on this book's cover, facing the Table of Contents and chapter 4, and on the last page of the Preface are by Nancy Nagle. These images are part of beautiful color galleries of yantras and mandalas. Both artists may be contacted by writing to me or by visiting their web pages:

www.nagledesign.com/ and
www.americansanskrit.com/inspire/yantras.html

The Arabic clock on this page is by Jan Abas (www.bangor.ac.uk/~mas009/). The swallowtail butterfly facing the Preface is by Jay Cossey (www.images.on.ca/JayC/).

I thank Harvey Heinz, Michael Keith, and Robert Stong for advice and extraordinary inspiration. Harvey Heinz has a wonderful set of web pages discussing magic squares. Although Harvey is a retired bookbinder and has no training in mathematics beyond high school, he is fascinated by number patterns and has been researching them since he was a teenager. For the last several years, he has made startling contributions to the field of magic stars. I also thank James Nesi, who familiarized me with the little-known magic squares of the late, great Fubine. The drawings on pages 374, 394, and 396 are by Paul Hartal, Ph.D. The drawing on page 405 is by April Pedersen.

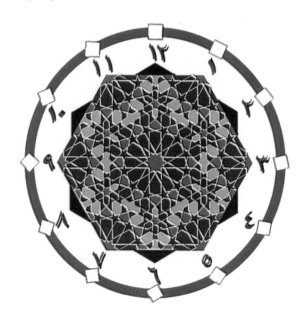

Introduction

Magic squares contain a lesson of great value in being a palpable instance of the symmetry of mathematics, throwing thereby a clear light upon the order that pervades the universe wherever we turn, in the infinitesimally small interrelations of atoms as well as in the immeasurable domain of the starry heavens.

–Paul Carus, in W. S. Andrews's *Magic Squares and Cubes*

What Is a Magic Square?

A *magic square* is a square matrix drawn as a checkerboard filled with numbers or letters in particular arrangements. Mathematicians are most interested in *arithmetic* squares consisting of N^2 boxes, called *cells*, filled with integers that are all different. Such an array of numbers is called a magic square if the sums of the numbers in the horizontal rows, vertical columns, and main diagonals are all equal. If the integers in a magic square are the consecutive numbers from 1 to N^2, the square is said to be of the Nth order, and the *magic number*, or sum of each row, is a constant symbolized as $\mathcal{S}$:

$$\mathcal{S} = \frac{N(N^2 + 1)}{(2)}$$

(The magic number is sometimes referred to as the *magic sum* or *magic constant*.) To derive this expression for $\mathcal{S}$, recall that the sum of the first m numbers in the arithmetic series $1 + 2 + 3 + \cdots + m$ is equal to $m(m + 1)/2$. In our case $m = N^2$ because we are interested in the sum of all numbers in the magic square. This means that the sum of the numbers in a magic square is $N^2(N^2 + 1)/2 = (N^4 + N^2)/2$. To get $\mathcal{S}$, we divide this result by N, which gives the sum for each of the N rows and N columns.

A few examples will help demystify these mathematical definitions. The simplest magic square possible is one of the third order, with 3×3 cells containing the integers 1 through 9, and with the magic sum 15 along the three rows, three columns, and two diagonals. In some sense, only one unique arrangement of digits, and its mirror image, is possible for a third-order square:

4	9	2
3	5	7
8	1	6

Third-order magic square

2	9	4
7	5	3
6	1	8

Mirror image

Here, $N = 3$ because there are three rows and three columns, and the magic sum $\mathcal{S}$ is 15 because the numbers in the rows, columns, and two diagonals sum to 15. For example, if you look at the square on the left, you will see that the sum of the numbers in the first row is $4 + 9 + 2 = 15$, the sum of the numbers in the first column is $4 + 3 + 8 = 15$, one of the diagonal sums is $4 + 5 + 6 = 15$, and so forth. We can also use the magic sum formula to compute the magic sum: $3(3^2 + 1)/2 = 15$. Notice that the mirror image is also a magic square. By rotating the square four times by 90 degrees, you can produce eight third-order magic squares.

To quickly understand why the number 5 must be at the center of the third-order square, consider all the ways in which the magic constant, 15, can be partitioned into a triplet of distinct positive integers:

$$9 + \mathbf{5} + 1, \quad 9 + 4 + 2, \quad 8 + 6 + 1, \quad 8 + \mathbf{5} + 2$$

$$8 + 4 + 3, \quad 7 + 6 + 2, \quad 7 + \mathbf{5} + 3, \quad 6 + \mathbf{5} + 4$$

In a third-order magic square, each of the three rows, three columns, and two diagonals must sum to 15. In other words, there must be eight sets of numbers that sum to 15, and these are represented by the eight previous sums. Since the center number belongs to one row, one column, and two diagonals, it must appear in four of the eight equations. The only such digit is 5. This means that 5 must be in the center of the 3×3 square.

Here is an example of a fourth-order magic square:

16	3	2	13
5	10	11	8
9	6	7	12
4	15	14	1

Fourth-order magic square

Cells placed at equal distances from the center cell and on opposite ends of an imaginary line through the center are called *skew-related cells.* A few examples of skew-related cells are shown in the following diagram. For example, the cells with the two smiley faces are skew related, the two alien faces are skew related, and so forth.

Skew-related cells

If we do not consider reflections and rotations, there is only one third-order magic square. As the order N increases,[1] the number of magic squares increases rapidly:

N	Number of Different Magic Squares
1	1
2	0
3	1
4	880
5	275,305,224

As far back as 1693, the 880 different fourth-order magic squares were published posthumously by French mathematician Bernard Frénicle de Bessy (1602–1675). Frénicle, one of the great magic square researchers of all time, was an eminent amateur mathematician working in Paris during the great period of French mathematics in the seventeenth century.[2] His magic square research was often carried out by a "method of exhaustion" rather than by deep mathematical theory, and his lists of solutions for fourth-order squares was published as an important part of a substantial treatise entitled *Des Quarrez ou Tables Magiques*, one of four treatises collected by Phillipe de la Hire (1640–1718), a French geometer who was also interested in magic square construction. In 1731, the treatise was republished as *Divers Ouvrages de Mathématique par Messieurs de l'Académie des Sciences* in the Hague, and it is this publication that is more accessible today.

It took humanity until 1973 to determine the exact number of order-5 squares—at which time Richard Schroeppel, a mathematician and computer programmer, used one hundred hours of PDP-10 computer time to arrive at this number.[3] Schroeppel prefers to divide this large number by 4 and give the total as 68,826,306 because, in addition to the eight variants obtained by rotation and reflection, there are four other variants that also preserve magical properties:

- Exchange the left and right border columns; then exchange the top and bottom border rows.

■ Exchange rows 1 and 2 and rows 4 and 5, then exchange columns 1 and 2 and columns 4 and 5.

As Martin Gardner points out,[4] when these two transformations are combined with the two reflections and four rotations, the result is $2 \times 4 \times 2 \times 2 = 32$ forms that can be called isomorphic (having essentially the same structure). With this definition of isomorphic, the number of "unique" fifth-order magic squares drops to 68,826,306.

Because you can always generate a magic square from an existing magic square by subtracting each integer in the square from $N^2 + 1$, the number of "unique" fifth-order squares might be decreased further. (The new square produced by this subtraction is sometimes called the magic square's *complement.*) When the center of a 5×5 square is 13, the complement is isomorphic with the original. If it is not 13, a different square results. If we broaden the term "isomorphic" to include complements, the count of fifth-order squares drops to about 35 million.

If the notion of a complement magic square leaves you confused, consider the following fourth-order magic square on the left. Here I have exchanged all the numbers x with new numbers $N^2 - x + 1$ to produce a new magic square on the right. The two squares are said to complement one another.

1	2	16	15
13	14	4	3
12	7	9	6
8	11	5	10

Fourth-order magic square

16	15	1	2
4	3	13	14
5	10	8	11
9	6	12	7

Complement magic square

Even though 880 fourth-order magic squares might seem like a lot, the magic is quite rare when one considers that there are roughly 2,615,348,736,000 possible arrangements of consecutive integers in a 4×4 array, where no integers are used more than once and rotations and reflections are not considered. (To compute this large number for any order N, use $N^2!/8$, where "!" is the mathematical symbol for factorial.)

As an example of a larger magic square, consider the following sixth-order square published in 1593 by Chinese mathematician Chêng Ta-wei:

27	29	2	4	13	36
9	11	20	22	31	18
32	25	7	3	21	23
14	16	34	30	12	5
28	6	15	17	26	19
1	24	33	35	8	10

Chêng Ta-wei square

Every row, column, and main diagonal adds up to 111. As preparation for the rest of the book, and for any reading you do in the magic square literature, let's review some informal definitions for various classes of diagonals using the Chêng Ta-wei square as an example:

- The *main* diagonals are the two diagonals that run from opposite corners of the square.
- The main diagonal from the top left corner cell to the bottom corner cell is called the *leading* or *left* diagonal.
- The main diagonal from the top right corner cell to the bottom corner cell is called the *right* diagonal.
- A diagonal that runs from one side of the square to the adjacent side and parallel to one of the main diagonals is called a *short diagonal*. In the Chêng Ta-wei square, one example of a short diagonal contains the numbers 32, 11, and 2. It is parallel to the right diagonal.
- Two short diagonals that contain the same number of cells and are on opposite sides of a main diagonal are called *opposite short diagonals*. The diagonal containing 9 and 29 and the diagonal containing 8 and 19 are opposite short diagonals.

- Two short diagonals that together contain N cells can form what is called a *broken diagonal.* One example is the cell containing 27 and the cells containing 24, 15, 30, 21, and 18.

I do not know if Chêng Ta-wei realized the fact that the four corners of his square add up to $\mathcal{S} \times 3/2$ or 74, as do the center four numbers and various skew-related pairs—for example, $32 + 5 + 14 + 23$ (highlighted by dark squares), $25 + 12 + 16 + 21$, etc. Most magic squares are like this one in the sense that their initially recognized symmetries and patterns give rise to a large number of other patterns that are not always known to the square's "author." It's as if there is a treasure chest of patterns waiting to be explored.

When consecutive integers 1 through N^2 are used, the resulting magic array of numbers is called a *pure magic square* or *traditional magic square.* In several chapters, I present unusual variations on this fundamental definition of magic square. I'll still refer to these imperfect beasts as "magic" even though they don't rigorously conform to the standard definition. To whet your appetite, consider some of the following examples. Instead of using consecutive integers starting with 1, we might use an arithmetic series starting with, say, 17 and with a difference of 3 between successive integers. A 4×4 square may be constructed[5] from the numbers 17, 20, 23, 26, 29, and so forth with magic constant $\mathcal{S} = 158$.

17	59	56	26
50	32	35	41
38	44	47	29
53	23	20	62

Imperfect magic square

Just as with traditional magic squares, a nice little formula exists to calculate the magic constant for these kinds of squares if you know the order N, the starting integer A, and the difference D between successive terms:

$$\mathcal{S} = N\left(\frac{2A + D(N^2 - 1)}{2}\right)$$

For the fourth-order magic square here, A is 17, D is 3, and N is 4. This yields $\mathcal{S} = 4 \times (34 + 45)/2 = 158$.

Imperfect magic squares can also be made from perfect magic squares simply by multiplying every number in a magic square by a constant. The results will be imperfect because the squares will not start with 1 and the numbers will not be consecutive.

There are also magic squares that are "more perfect" than perfect magic squares because they satisfy all kinds of wonderful criteria beyond the minimum of having the sums of integers in each row, column, and main diagonal equal to a constant. Various examples of these "hyper-magical" squares are discussed throughout this book.

A Brief History of Magic Squares

Lo-Shu

Some of the third- and fourth-order magic squares in the previous section were discovered centuries ago in India and China, where magic squares engraved on stones or metals were worn as charms to protect the wearer from evil and to bring good fortune. Most scholars believe the magic square originated in China and was first mentioned in a manuscript from the time of Emperor Yu, around 2200 B.C. The first square had $3 \times 3 = 9$ cells, each with Chinese characters equivalent to 1 through 9 (Figure 1).

4	9	2
3	5	7
8	1	6

Yu magic square

Although historians trace references to the Yu magic square back no further than the fourth century B.C., there is a strange legend[6] that has Emperor Yu discovering the magic square while walking along the Lo River (or Yellow River), where he saw a mystical turtle crawling on the river bank. I think we can take most of

Figure 1
The first known magic square from around 2200 B.C. The Chinese characters stand for the numbers 1 through 9.

this story with a bit of skepticism, but the legend is that the turtle was ordinary in all respects except that its shell had a series of dots within squares. To Yu's amazement, each row of squares contained fifteen dots, as did the columns and diagonals. As he studied the turtle shell further, he also found that when he added any two cells directly opposite along a line through the center square, such as 2 and 8, he always got the sum of 10.

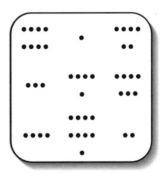

A piece of Yu's turtle shell

Emperor Yu had the turtle taken back to his palace for further study, and news of the mystical turtle began to spread to the nearby villages and eventually to other countries.

The turtle spent the rest of its easy life in Yu's court, and it became the most famous turtle in the world, having the company of famous mathematicians, kings, and international visitors. Soon the Lo-shu, as the pattern was later called, began appearing on

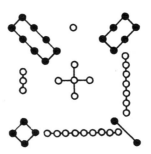

Figure 2
An ancient representation of the Lo-shu. Count the number of dots in each figure to form a 3 × 3 magic square.

charms and magic stones as it does to this day. In these charms, the arrangements of dots is usually depicted as in Figure 2, where the even numbers are represented with little filled circles and the odd numbers with open circles. (Have you ever noticed that cruise ships often feature the Lo-shu on their main deck as a pattern for the game of shuffleboard?)

From the fourth century B.C. until the tenth century, the pattern was a mystical Chinese symbol of great importance. The even numbers were identified with yin, the female principle, and the odd numbers with yang, the male principle. The central 5 represented the earth, surrounded by evenly balanced four elements in yin and yang: 4 and 9 represented metal, 2 and 7 fire, 1 and 6 water, and 3 and 8 wood.

4 (Metal)	9 (Metal)	2 (Fire)
3 (Wood)	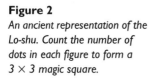 (Earth)	7 (Fire)
8 (Wood)	1 (Water)	6 (Water)

Chinese interpretation

For centuries after their interpretation as magic squares, these squares were crucial elements of Chinese numerology, used in imperial rituals by necromancers casting spells and as the basis for prophecies and horoscopes. These kinds of squares were introduced into Europe sometime during the first millennium A.D. The first known writer on the subject was Emanuel Moschopoulus, a Greek who lived in Constantinople around A.D. 1300. He is believed to have discovered two methods for constructing magic squares.

This same square had importance in the Islamic tradition because it was believed to contain the nine letters that were revealed to Adam, that is, the first nine letters of the Arabic alphabet in the old Semitic sequence (which is used to this day when letters are taken as numbers). The even numbers in the corners are read according to their numerical value as *buduh*, and this word, sometimes interpreted as the name of a spirit, often appears on the walls to protect a building or on amulets worn around the neck or on the upper arm.

I am most fascinated by the Lo-shu square because of its ubiquity. We find it venerated by civilizations of almost every period and continent. The Mayan Indians of southern Mexico and Central America were fascinated by it, and today it is used by the Hausa people of northwestern Nigeria and southern Niger as a calculating device with mystical associations. The square was respected by the ancient Babylonians and was used as a cosmic symbol in prehistoric cave drawings in northern France.[7] In Islam, it symbolized the power of Allah's omnipresence. Members of secret societies used it as a code by drawing lines through the cells containing particular numbers in order to form secret symbols.

Geber, the eighth-century Islamic alchemist (or a later writer using his name), believed he had found the key to the elements in this simple square. He divided the square into two parts, as shown next, with the four numbers in the bottom left-hand corner adding up to 17, which was regarded as auspicious in many cultures. For Christians, 17 symbolizes the star of the Magi. For kabbalists, the seventeenth path leads to reward for the righteous. Geber noticed that the remaining numbers of the square add up to 28, the number of letters in the Arabic alphabet.

4	9	2
3	5	7
8	1	6

Geber's square

Geber had some pretty odd ideas. One of his goals was to turn cheap metals into gold using elixirs ranging from "love-in-a-mist" to gazelle urine. By numbering various substances according to the numbers in his magic square, he thought he could determine how to turn the metal into gold. Geber also arranged the numbers around the central 5 in different ways for different magical properties. For example, the fire square and the earth square are represented as follows.

6	1	8
7	5	3
2	9	4

Fire square

2	7	6
9	5	1
4	3	8

Earth square

Magic squares such as these represented various objects in the solar system. Cornelius Agrippa (1486–1535)–physician, astrologer, and Catholic theologian–constructed squares of orders 3, 4, 5, 6, 7, and 9, which he associated with the seven known (astrological) "planets": Saturn, Jupiter, Mars, the Sun, Venus, Mercury, and the Moon. Agrippa had a colorful life that included various dangerous run-ins with the Church and jobs as an occult scholar, lawyer, and military strategist. Agrippa's *De Occulta Philosophia* stimulated Renaissance study of magic and got his name into early Faust legends. Agrippa believed that a magic square containing the digit 1–which exhibits the magic constant of 1 in all directions–represented God's eternal perfection.

God's magic square

Agrippa and his colleagues considered the sad discovery that a 2×2 magic square could not be constructed as proof of the imperfection of the four elements: air, earth, fire, and water. Others believed that the nonexistence of a 2×2 magic square resulted from humans' original sin.

Figure 3 shows a Jupiter amulet with magic sum 34. If this configuration is engraved on a silver tablet during the time that the planet Jupiter is ruling, it is supposed to produce wealth, peace, and harmony.[8] Figure 4 shows a Mars amulet with the magic sum 65. If engraved on an iron plate or a sword when the planet Mars is in the ruling position, this amulet is said to bring success in lawsuits and victory over the owner's enemies.

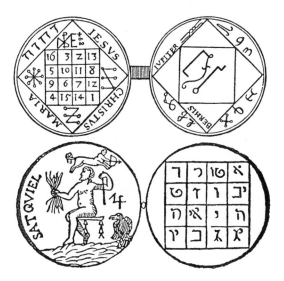

Figure 3
Jupiter amulet with magic sum 34. If this magic square is engraved on a silver tablet during the time the planet Jupiter is ruling, it is supposed to produce wealth, peace, and harmony.

Figure 4
Mars amulet with magic sum 65.

The Arabs believed magic squares had great powers in all aspects of life. Some squares were used to protect and help lame children. The Arabs even showed certain squares to women in labor and then placed them over their wombs to facilitate births. Magic squares were also written or embroidered on the shirts of soldiers, mainly in Turkish and Indian areas. According to Annemarie Schimmel,[9] author of *The Mystery of Numbers*, people believed that such shirts had to be made by forty innocent virgins in order to work!

In Islam, the number 66 corresponds to the numerical value of the word "*Allah.*" Figure 5 is an Islamic magic square that expresses the number 66 in every direction when the letters are converted to numbers. The square's grid is formed by the letters in the word "*Allah.*" Magic squares such as this were quite common in the Islamic tradition but did not reach the West until the fifteenth century.

Quick Philosophizing

Before proceeding, let us pause and reflect upon some of these early magic squares. Why is it that Chinese emperors and empresses,

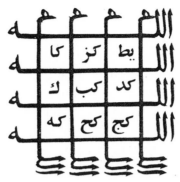

Figure 5
*An Islamic magic square that expresses
the number 66 in every direction. The
grid is formed by the letters in the word
"Allah," whose numerical value is also 66
(nineteenth century, Damascus).*

Babylonian astrologer-priests, prehistoric cavemen in France, ancient Mayans of the Yucatan, and modern Hausa tribesmen were all convinced that the Lo-shu square held the secret of the universe?[10] Could they have all learned of this number arrangement from a single primal source, or is it more likely they stumbled across it independently?

The omnipresence of the Lo-shu reinforces the idea that mathematics and mysticism have fascinated humanity since the dawn of civilization. Throughout history, number arrays held certain powers that made it possible for mortals to seek help from spirits, perform witchcraft, and make prayers more potent. Numbers have been used to predict the end of the world, raise the dead, find love, and prepare for war. Even today, serious mathematicians sometimes resort to mystical or religious reasoning when trying to convey the power of mathematics.

Has humanity's long-time fascination with mathematics arisen because the universe is constructed from a mathematical fabric? In 1623, Galileo Galilei reinforced this belief by stating his credo: "Nature's great book is written in mathematical symbols." Plato's doctrine was that God is a geometer, and Sir James Jeans believed God experimented with arithmetic. Newton supposed that the planets were originally thrown into orbit by God, but even after God decreed the law of gravitation, the planets required continual adjustments to their orbits.

What does God think of magic squares? Certainly the world, the universe, and nature can be reliably understood using mathematics. Nature *is* mathematics. The arrangement of seeds in a sunflower can be understood using Fibonacci numbers (1, 1, 2, 3, 5, 8, 13, . . .), named after the Italian merchant Leonardo Fibonacci of Pisa. Except for the first two numbers, every number in the sequence equals the sum of the two previous. Sunflower heads, like other flowers, contain two families of interlaced spirals—one winding clockwise, the other counterclockwise. The numbers of seeds and petals are almost always Fibonacci numbers.

The shape assumed by a delicate spider web suspended from fixed points or the cross section of sails bellying in the wind is a catenary—a simple curve defined by a simple formula. Seashells, animal's horns, and the cochlea of the ear are logarithmic spirals that can be generated using a mathematical constant known as the golden ratio. Mountains and the branching patterns of blood vessels and plants are fractals, a class of shapes that exhibit similar structures at different magnifications. Einstein's $E = mc^2$ defines the fundamental relationship between energy, matter, and the speed of light. And a few simple constants—the gravitational constant, Planck's constant, and the speed of light—control the destiny of the universe. I carry this reasoning further in my book *The Loom of God,* where I state, "I do not know if God is a mathematician, but mathematics is the loom upon which God weaves the fabric of the universe."[11]

India, Germany, and Beyond

It was not until the first millennium A.D. that a fourth-order square made its appearance in society. In fact, the first record of a fourth-order square appeared in a Jaina inscription hanging over the gate at Klajuraho, India, in A.D. 1100. (Jainism is a religion and philosophy of India, founded in the sixth century B.C. by Vardhamana.) This beautiful array of numbers appears to show an advanced knowledge of magic squares because all the broken diagonals have numbers that also sum to the magic constant 34.

7	12	1	14
2	13	8	11
16	3	10	5
9	6	15	4

Jaina magic square

For example, consider the broken diagonal $7 + 6 + 10 + 11 = 34$. These kinds of squares are also known as *diabolic*, a term discussed in chapter 2. The Jaina square contains other interesting patterns. For example, we can sum the top two rows and the bottom two rows, and the right two columns and the left two columns, revealing additional symmetries:

9	25	9	25
25	9	25	9

Row clusters

19	15	19	15
15	19	15	19

Column clusters

Today, number arrays are still used in India, drawn on paper or engraved on metal. Some squares are even regarded as being the personification of God. As pointed out by Richard Webster in *Numerology Magic,* by constructing a magic square, the individual is thought to be communicating directly with the universe's "life-force," and therefore various wishes can be granted, including the curing of piles, the alleviation of pain during childbirth, or the annoyance of an enemy. However, not all of the India magic squares are magic from a numerical point of view or contain magic sums. Consider the following square that a woman may use when searching for a husband. The numbers are drawn on a china plate with a crayon and then washed off the plate with water that the woman drinks:

24,762	24,768	24,771	25,320
24,770	24,758	24,763	25,341
24,759	24,773	24,766	25,325
24,767	24,761	24,760	25,344

Finding the perfect husband

Whenever possible, this number array should be written with a special ink known as *Ashat Gandh.* This is a mixture of several items, the most important of which is water from the Ganges River. I would be interested in hearing from readers who ascertain any significance in the use of these particular numbers.

There is even a special number array for people who feel that others are talking behind their backs. The following array, known as *Swara-e-Manafakoon,* can also be used if people have hurt you in some way. The 4×3 rectangle contains numbers of three, four, and five digits, and one of the numbers is repeated. Don't get mad, get even! If you'd like to try this square, you are supposed to recite the numbers 160 times and then keep the number array with you to prevent a recurrence of the problem.

7753	177	408	7355
7754	157	1530	17,750
17,749	177	156	17,751

Swara-e-Manafakoon

The following 3×3 square (a version of the Lo-shu) is used in India to find missing people. Written when someone vanishes without a trace, the array is hung from a tree and is thought to draw the person back home. Other 3×3 squares are used to create harmony between a man and a woman. One such square should be drawn on a Wednesday or Friday, and both people should keep a copy with them.

6	7	2
1	5	9
8	3	4

Find missing people

My favorite centuries-old European magic square is Albrecht Dürer's, which is drawn in the upper right-hand column of his etching *Melencolia I* (Figure 6). Dürer, the greatest German Renaissance artist, included a variety of small details in the etching that have confounded scholars for centuries. We seem to see the figure of a brooding genius sitting amid her uncompleted tasks. There are scattered tools, flowing sands in the glass, the magic square beneath the bell, and the swaying balance. Scholars believe that the etching shows the insufficiency of human knowledge in attaining heavenly wisdom or in penetrating the secrets of nature. As I mentioned, Renaissance astrologers linked fourth-order magic squares to Jupiter, and these squares were believed to combat melancholy (which was Saturnian in origin). Perhaps this explains the square in Dürer's engraving. Dürer's 4×4 magic square is represented as

16	3	2	13
5	10	11	8
9	6	7	12
4	15	14	1

Dürer magic square

The square contains the first 16 numbers and has some fascinating properties. The two central numbers in the bottom row read "1514," the year Dürer made the etching. Scholars wonder if "1514" appeared accidentally or if Dürer constructed it intentionally. The rows, columns, and main diagonals sum to 34. In addition, 34 is the sum of the numbers of the corner squares $(16 + 13 + 4 + 1)$ and of the central

Figure 6
Melencolia I, by Albrecht Dürer (1514). This figure is usually considered the most complex of Dürer's works, the various symbolic nuances confounding scholars for centuries. Why do you think he placed a magic square in the upper right? Scholars believe that the etching shows the insufficiency of human knowledge in attaining heavenly wisdom or in penetrating the secrets of nature.

2×2 square $(10 + 11 + 6 + 7)$. The sum of the remaining numbers is $68 = 2 \times 34$. Figure 7 summarizes all the amazing "34" sums from adding other configurations of cells. Just sum the numbers connected by lines. For clarity, dots represent the numbers to be added.

As you will learn in chapters 1 and 2, the Dürer square is an associated square of doubly even order because the sum of any two skew-related cells are the same. Don't worry about this termi-

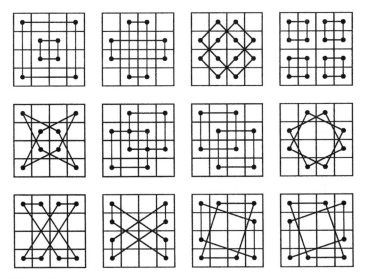

Figure 7

The Dürer square yields a magic sum in an amazing number of ways in addition to the traditional ways. The patterns, although reminiscent of the symmetries of certain crystals, tell us a good deal about the properties of this square. Try adding any of these cell configurations, where connected dots represent numbers to be added.

nology right now; the fun has just begun and doesn't require sophisticated words. For example, the sum of the squares of the integers in the first and second rows equals the sum of the squares of the integers in the second and fourth rows, equals the sum of the squares of the integers in the first and third rows, equals the sum of the squares of the integers in the third and fourth rows, equals the sum of the squares of the integers not in the main diagonals, equals 768! All of this is also true for the columns. The Dürer square contains other interesting patterns. For example, like the Jaina square, we can sum the top two rows and the bottom two rows, and the right two columns and the left two columns, revealing additional symmetries:

21	13	13	21
13	21	21	13

Row clusters

19	15	15	19
15	19	19	15

Column clusters

Even better, the sum of the cubes of the integers in the main diagonals equals the sum of the cubes of the integers not in the main diagonal.

256	9	4	169
25	100	121	64
81	36	49	144
16	225	196	1

Dürer squared

4096	27	89	2197
125	1000	1331	512
729	216	343	1728
64	3375	2744	1

Dürer cubed

I wonder if Dürer thought there was something profound about the magic sum 34? Probably not, but don't you wonder why it is even possible to create such a square with all the marvelous symmetries? Additional features of the Dürer square are discussed in chapter 3. For now, I will whet your appetite by pointing out yet another symmetry, simply by connecting the odd and even numbers (Figure 8).

It is also possible to use bicolored tiles to highlight additional symmetry and beauty. For example, I have used a light-to-dark tile to represent odd numbers and a dark-to-light tile to represent even squares:

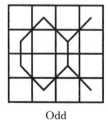

Odd

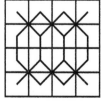

Even and odd

Figure 8
Connecting odd and even numbers in the Dürer square.

Dürer magic tile

Try this coloring method on other magic squares of your own design.

Templar Magic Square

The templar magic square is a particular arrangement of the words in the Latin sentence "*Sator Arepo tenet opera rotas,*" which has been translated variously as "The farmer Arepo keeps the world rolling," "Farmer Arepo keeps the wheels at work," "Farmer Arepo steers the plough," "The sower Arepo works with the help of a wheel," or "Arepo the farmer holds the works in motion." Arranged both vertically and horizontally, the seemingly unimportant word TENET forms the two arms of a central cross. This first-century square has been found in excavations of ancient Pompeii, Italy. The square continued to be used during the nineteenth century in Europe and the United States for protection against fire, sickness, and other disasters.

S	A	T	O	R
A	R	E	P	O
T	E	N	E	T
O	P	E	R	A
R	O	T	A	S

Templar magic square

In Rome during the Middle Ages this square was inscribed on a variety of common, everyday objects, such as utensils and drinking vessels. It was also placed above doorways. People believed that the square had magical properties and that making it visible would ward off evil spirits. I am not sure why there are conflicting translations of the Latin. For example, some have translated the words in this square less literally, as in "The Creator (or Savior) holds the working of the spheres in his hands," or "The sower [i.e., God] in his field controls the workings of his tools [i.e., us]." Perhaps it is these more expansive interpretations that have given the square meaning through the ages.

Most authors who have reported on this square in the past have not noticed that the following hidden message arises if the square's contents are rearranged:

```
                    P
                    A
        A           T           O
                    E
                    R
    P   A   T   E   R   N   O   S   T   E   R
                    O
                    S
        O           T           A
                    E
                    R
```

This translates to "Our Father, Our Father" in the shape of a cross, which must have been a potent Christian message posted near doorways to let Christians in the neighborhood know that sympathetic families were nearby. The four remaining letters may represent alpha and omega—the beginning and the end. Note that this hidden message is an example of what codebreakers call steganography, an encryption technique that conceals a message so that the casual viewer doesn't even know there is a secret message waiting to be found.

The French Connection

Humanity did not know very much about magic squares until the seventeenth century, when there was an explosion of interest in France by mathematicians such as Simon de la Loubère, Claude Gaspar Bachet de Méziriac, Philippe de la Hire, and others discussed in chapter 1. In 1838, Jean-Yves Violle published three volumes on magic squares that contained hundreds of illustrations. It was during this period that the 880 different fourth-order magic squares were found, not counting rotations and reflections.

Although research in magic squares still continues unabated today, some researchers in the early 1900s seemed to think that there was not much more to know about these impressive number patterns. For example, world-renowned puzzlist Henry E. Dudeney (1857–1930) wrote in his 1917 book *Amusements in Mathematics*,

> Of recent years many ingenious methods have been devised for the construction of magics (magic squares), and the law of their formation is so well understood that all the ancient mystery has evaporated and there is no longer any difficulty in making squares of any dimensions. Almost the last word has been said on this subject.[12]

Dudeney's assertions proved to be not true. In fact, he quickly invalidated his own prediction when he continued to write about magic squares. However, Dudeney did not think there was much value in classifying magic squares. Dudeney wrote, "A man once said that he divides the human race into two great classes: those that take snuff and those who do not. I am not sure that some of our classifications of magic squares are not almost as valueless."[13] Coincidentally, one of the world's most important, thorough, and beautiful books on magic squares, W. S. Andrews's *Magic Squares and Cubes,* was also published in 1917, at about the same time Dudeney was making these sorts of statements.

To my mind, we have certainly made remarkable progress, and the field of magic square study is wide open. Consider that *Homo sapiens* took about 20,000 years to discover the Yu magic square (starting from our origins as anatomically modern *Homo sapiens* in the Upper Paleolithic period). It took another 3500 years for magic squares to permeate Europe, thanks in part to Emanuel Moschopoulus. About 390 years after that, Bernard Frénicle de Bessy published the 880 different fourth-order magic squares. Then, 280 years later, Richard Schroeppel determined all order-5 squares. About 20 years later, John Hendricks discovered four-, five-, and six-dimensional magic hypercubes. If this isn't progress, nothing is! What do you think comes next?

Geometrical Diagrams

Throughout this book I have drawn *geometrical diagrams* that reveal some of the startling symmetries of magic squares not obvious from the number arrays. The traditional method for drawing such diagrams, which probably dates back more than a century,[14] lists all of the numbers in the square and connects those numbers that are in the same rows. Usually the geometrical diagrams are presented in two columns of interconnected dots, and you'll see some of my favorite diagrams in Gallery 1, which includes information on the Franklin square. As I mention in Gallery 1, I like to meditate upon the geometrical diagrams as if they are mandalas–Hindu and Buddhist symbolic diagrams used in the performance of sacred rites and as an instrument of meditation. Traditionally, mandalas were painted on paper or cloth, drawn on a carefully prepared ground with white and colored threads or with rice powders, fashioned in bronze, or built in stone. Perhaps mandalas of the twenty-first century will include the geometric diagrams for huge magic squares with dimensions greater than 1000×1000.

The diagrams are not just beautiful to look at but also have a practical side because they give us hints about the existence of higher-dimensional magic figures. They also suggest regular laws in the arrangement of squares that persist from low- to high-order squares.

To whet your appetite, here is an 8×8 magic square,[15] with magic constant 260, and its geometrical diagram (Figure 9). (Incidentally, notice that skew-related cells add up to the same sum, for example, $29 + 36 = 8 + 57 = 33 + 32 = 65$.) To help you find your way, take a look at the first row in the magic square (1, 7, 59, 60, 61, 62, 2, 8) and see how it is represented by the line that intersects these numbers in the upper left part of the geometrical diagram.

1	7	59	60	61	62	2	8
16	10	54	53	52	51	15	9
48	47	19	21	20	22	42	41
33	34	30	28	29	27	39	40
25	26	38	36	37	35	31	32
24	23	43	45	44	46	18	17
56	50	14	13	12	11	55	49
57	63	3	4	5	6	51	64

Eighth-order magic square

The geometrical diagram makes it clear that there are two different underlying symmetries that characterize the order-8 square. Gallery 1 shows an order-8 square with *three* underlying symmetries.

The best part about these diagrams is that the structures do not seem random. Symmetrical themes are always repeated, like movements in a symphony or atoms in a crystal.

I would be interested in hearing from readers who have invented other kinds of geometrical diagrams. For example, the diagrams might be modified to form circular geometrical diagrams to reveal hidden structure and order. This approach is similar to the traditional two-column interconnected dots approach that connects numbers in the same magic square row, but instead you

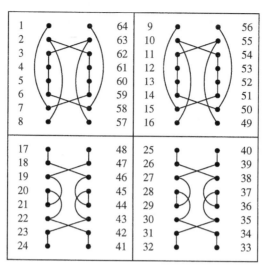

Figure 9
Geometrical diagram for an 8 × 8 magic square.

can abolish the dot columns in favor of a more general diagram that positions consecutive integers around the circumference of a circle. These circular (or other kinds of) geometrical diagrams might be quite convenient because they fit on one page, even for huge magic squares. The traditional geometrical diagram can also be made to fit on a single page by chopping it into subsections.

Zen Mathematics

The Gohonzon of the Soka Gakkai

At that time in the Buddha's presence, there was a tower adorned with the 7 treasures, 500 yojanas in height and 250 yojanas in width and depth. . . . It had 5000 railings, 10,000 rooms, . . . 10,000,000,000 jeweled bells.

–The Lotus Sutra

An elephant is contained in a blade of grass.

–Ancient Buddhist saying

In previous sections, I've discussed how magic squares were used in magic, divination, alchemy, and other mysterious settings. The use of mathematical and geometrical patterns for magic and meditation has a rich history. In Japan, there exists a religious group, the members of which meditate using visual aids resembling fractals and other ornate geometries. *Soka Gakkai* is an association of laypeople who practice Nichiren-sho-shu Buddhism. The association was founded in Japan in 1937; however, the practice of Nichiren-sho-shu Buddhism dates back centuries to the followers of the Japanese Buddhist saint Nichiren (1222–1282), an ardent nationalist who attacked the established contemporary religious and political institutions of Japan. After exploring different facets of Buddhist scripture, Nichiren concluded that only one scripture gave the truth–the *Lotus Sutra* (the scripture of the Lotus of the Good Law).

One way the Soka Gakkai pay devotion to the Lotus Sutra is through a ritual drawing, or mandala, called the *gohonzon*. The gohonzon contains the names of divinities mentioned in the sutra arranged around the name of the Lotus Sutra. In particular, the gohonzon contains the sacred phrase *Nam-myoho Renge-kyo* (Devotion to the Lotus Sutra) written vertically in the center, around which are arranged names of various deities–perhaps to safeguard the seven-ideograph phrase. Today, millions of followers chant the phrase thousands of times a day.

The gohonzon mandala, an example of which is shown in Figure 10, is placed on an altar in Nichiren temples, as well as in the homes of Nichiren Buddhists. The original *dai gohonzon*–the super-gohonzon for all humanity–resides in a temple in Kyoto. No photographs of the dai gohonzon are allowed, and no pictures have been published. To the Nichiren Buddhists, the gohonzon symbolizes the superiority of the Lotus Sutra over other religions and sects. Nichiren himself drew many gohonzon mandalas that he gave to his followers. Since the followers were only entitled to the mandalas while they lived, the mandalas were returned to the temples after the deaths of the followers.

The gohonzon reminds me of a fractal; all the different renditions I have seen have visual details on at least two different size scales. When I talked to Gary Adamson, expert and practitioner of Nichiren-sho-shu Buddhism, he indicated that the gohonzon mandala represents the

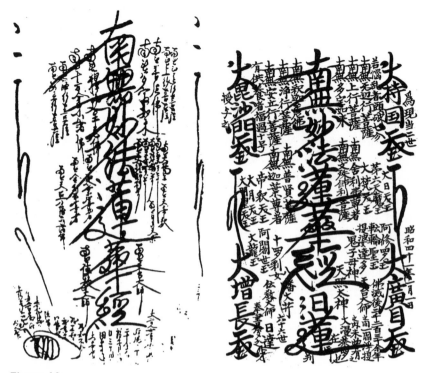

Figure 10
Two gohonzons of Nichiren (thirteenth century).

Cosmos, the totality of existence, or, in modern mathematical terms, "the class of all sets." Further, he believes that the gohonzon's basic theme is "the whole is greater than the sum of its parts." Gary says, "When people chant to the gohonzon, they are not looking at individual parts of the fractal figure, but the complete whole." Gary sent me the ancient gohonzons in Figure 10 from the Nichiren Shoshu sect. The world of Buddahood resides at the top of the figure, and hell is below. Buddhism has a classification of ten dimensions, the topmost dimension being that of enlightened Buddhas.

Nichiren Buddhism contains various numerical curiosities—just as do many mystical groups or religions mentioned in this book in relation to magic squares. For example, one of the main elements of Nichiren's teachings is the theory of *ichinen sanzen* (1 thought, 3000) which means that the mind, at any moment, contains 3000 different aspects of the universe. To arrive at the figure 3000, Buddhists consider the human mind to be in one of ten states at any particular moment: hell, craving, animalness, anger, tranquility, rapture, intellectual pleasure, truth, mercy and wisdom, and, finally, Buddhahood. In each state of mind there are the same ten aspects, bringing the total number of states of mind to 100. There are an additional ten factors of existence–appearance, nature, entity, power, action, cause, relationship, effect, reward, and consistency–which bring the 100 states to 1000. Finally, the 1000 is multiplied by 3 to take account of the three categories of human life: physical and mental abilities, the relationship between the individual and other members of his or her community, and the environment in which an individual personality exists.

Makiguchi Tsunesaburo, the founder of Nichiren-sho-shu Buddhism, suggests that the goals of life are beauty, gain, and goodness. Through Nichiren Buddhism, an individual may attain these goals.

Gary Adamson indicates that the gohonzon is unlike other popular visual meditation aids, such as magic squares or the shri-yantra of India, because the gohonzon is the most free-flowing of all mandalas, having a random component along with a beautiful structure "analogous to many natural phenomena like twigs of a tree." The shri-yantra, an ancient Vedic symbol that represents the seed of the universe–a place that exists beyond space and time–is purely geometric, using triangles, squares, circles, and parallelograms (Figure 11). Some medieval and later Hindu temples in India contain shrines with shri-yantra engravings. The shri-yantra is also engraved on foil and placed in a metallic case worn as an amulet for health.

The shri-yantra is ripe for magic-sum exploration. For example, can you put consecutive numbers at each corner and intersection of the triangles so that the numbers along each line segment produce the same magic sum? So far no one has been able to create such a labeling of the points on the figure.

The mouse child uses the image as a yantra to meditate on nothing and infinity.

–*Daedalus,* 1980

Figure 11
The shri-yantra of India, a geometrical diagram used as an aid to meditation. This yantra is composed of nine juxtaposed triangles arranged to produce forty-three small triangles. Four of the nine primary triangles point upward, representing male cosmic energy, and five point downward, symbolizing female cosmic power. In southern India, the shri-yantra is an object of worship.

Hermetic Geometry

We who are heirs to three recent centuries of scientific development can hardly imagine a state of mind in which many mathematical objects were regarded as symbols of spiritual truths.

–Philip J. Davis and Reuben Hersh, *The Mathematical Experience*

Completely involved as he was in Hermetism, Bruno could not conceive of a philosophy of nature, of number, of geometry, of a diagram, without infusing into these divine meanings. Bruno based memory on celestial images which are shadows of ideas in the soul of the world, and thus unified the innumerable individuals in the world and all the contents of memory.

–Frances Yates, *Giordano Bruno and the Hermetic Tradition*

Through the ages, various patterns, such as in magic squares or in the gohonzon, have been created to induce spiritual forces to influence material forces. Perhaps the most geometrical of these mystical figures comes from Hermetic[16] geometry, where the diagrams represent pure celestial forms. The design on paper was supposed to induce a resonance with its celestial counterpart, and, as a result, the figure was thought to have various powers. For example, the symmetrical patterns were used to achieve

Figure 12
*The Hermetic figures of Giordano Bruno (1588): (a) Figura Mentis, (b) Figura Intellectus,
(c) Figura Amoris, (d) Zoemetra.*

personal gain, cure diseases, find love, or harm one's enemies. Figure 12
shows Hermetic designs from 1588 by Giordano Bruno. The Hermetic
geometers thought that these designs were keys to the universe.

Giordano Bruno (1548–1600) was an Italian philosopher, mathematician, astronomer, and occultist. Some of his theories antici-
pated modern science. For example, he rejected the traditional
Earth-centered universe and believed in a multiplicity of worlds. He
also believed that the Bible should be followed for its moral teaching
but not for its astronomical implications.

Giordano Bruno's patterns are of particular interest because of their geometrical and (often) recursive shapes. Occasionally, Bruno adds objects such as serpents and lutes. If you look closely, you can see that Figure 12c ("Figura Amoris") actually has the word "Magic" written in the diagram.

Bruno had a dramatic life. He thought himself a messiah, which was not uncommon for magicians of the Renaissance. The Venetian Inquisition was not pleased with Bruno's ideas and forced him to justify them. He gave the Inquisition a detailed technical account of his philosophy as if he were lecturing great scholars. He told them that he believed the universe to be infinite, because the infinite divine power would not produce a finite world. The Earth was a star, as Pythagoras thought, like the moon, other planets, and other worlds that were infinite in number.

At the end of the Venetian trial, Bruno fully recanted various heresies of which he was accused and threw himself at the mercy of the judges. By law, his case was sent on to Rome, where he was imprisoned as the trial dragged on for years. Finally, in 1599, a judge listed eight heresies taken from the works of Bruno that Bruno was required to repudiate. Bruno countered that there was nothing in his works that was heretical and that the church was merely misinterpreting them. Bruno finally declared that he did not even know what he was expected to retract. As a result, Bruno was sentenced as an impenitent heretic. His death sentence was read, and Bruno addressed his judges, "Perhaps your fear in passing judgment on me is greater than mine in receiving it." He was gagged and burned alive in Rome on February 17, 1600.

Unfortunately, today we do not have the report of the Venetian Inquisition, and so we do not know the eight heretical propositions that Bruno was required to recant. Scholars believe they have something to do with God's infinity implying an infinite universe, the mode of creation of the human soul, the motion of the Earth, the soul, the multiplicity of worlds, and the desirability of using magic. Other problems arose from the fact that Bruno believed that Moses performed his miracles by magic and that Christ was a magician.

Frances Yates, writing in *Giordano Bruno and the Hermetic Tradition*, notes

> The Renaissance magic was turning towards number as a possible key to operations, and the subsequent history of man's achievements in applied science has shown that number is indeed a master-key, or one of the master-keys, to operations by which the forces of the cosmos are made to work in man's service. However, neither Pythagorean number, organically wedded to symbolism and mysticism, nor Kabalistic conjuring with numbers in relation to the mystical powers of the Hebrew alphabet, will of themselves lead to the mathematics which really work in applied sciences.[17]

Other, more recent mystics have also used geometrical diagrams in spiritual realms. For example, Figure 13 shows the Sahasra chakra, or energy centers, of Tantric literature. This diagram was drawn in the mid-1800s by Yogi Lahiri Mahasaya (1828–1895).

I can easily imagine future generations, or even alien civilizations, that place great faith in magic squares, like some of our ancients did. In our modern era, God and mathematics are usually placed in totally separate arenas of human thought. But this has not always been the case, and even today many mathematicians find the exploration of mathematics akin to a spiritual journey. The line between religion and mathematics becomes indistinct. In the past, the intertwining of religion and mathematics has produced useful results and spurred new areas of scientific thought. Consider, as just one small example, numerical calendar systems first developed to keep track of religious rituals. Mathematics, in turn, has affected religion because mathematical reasoning and proofs have contributed to the development of theology.

In many ways, the mathematical quest to understand infinity and higher-dimensional magic squares parallels mystical attempts to understand God. Both religion and mathematics attempt to express relationships between humans, the universe, and infinity. Both have arcane symbols and rituals and impenetrable language. Both exercise the deep recesses of our minds and stimulate our imagination. Mathematicians, like priests, seek ideal, immutable, nonmaterial truths and then often try to apply these truths in the real world.

যোগিরাজ শ্রীশ্রীশ্যামাচরণ লাহিড়ী মহাশয়ের প্রদত্ত
ষট্‌চক্র চিত্র ও ৪৯ বায়ুর বিবরণ।

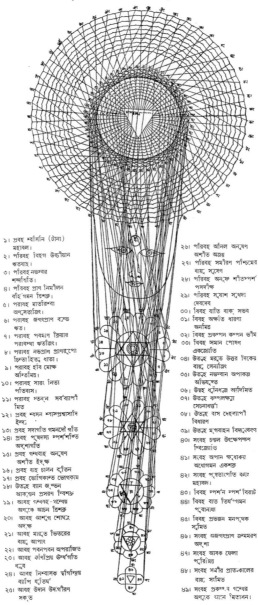

১। প্রবহ শর্ব্বাসন (টানা) মহাবল।
২। পরিবহ বিহগ উড্ডীয়ান ক্ষতবাহ।
৩। পরিবহ নভচর শর্ব্বাহিত।
৪। পরিবহ প্রাণ নিমৌলন বাহ্যগমন বিশঙ্ক।
৫। পরিবহ মাতারিশ্বা অনুসন্তাপাং।
৬। পরিবহ জগপ্রাণ বম্ভ ঋত।
৭। পরিবহ পবমাণ ক্রিয়ার পরাবস্থা ক্ষতাঃাং।
৮। পরিবহ নভপ্রাণ প্রাদর্পে চিন্তাহিতু ধাতা।
৯। পরিবহ হাবি মোক্ষ আত্মামিহ।
১০। পরিবহ সারং নিতা পাতবাস।
১১। পরিবহ স্তনন সর্ব্বব্যাপী মিত।
১২। প্রবহ শ্বসন শ্বাসপ্রখাসাদী ইন্দু।
১৩। প্রবহ সদাগতি গমনেঽ গতি।
১৪। প্রবহ পুষ্পদস্য চপ্পর্শশিন্ত অদ্ধশাগতি।
১৫। প্রবহ গন্ধবাহ অনুষ্ণ অশীত ইদৃক।
১৬। প্রবহ বায় চালন ব্যতান।
১৭। প্রবহ ভোগিকান্ত ভোলগকাম।
১৮। উতরহ বান জৃম্ভন আকুঞ্চন প্রসরণ বিশস্ত।
১৯। আবহ গন্ধবহ-গন্ধের্য অন্যেক আচন ত্রিশঙ্ক।
২০। আবহ আশগে শোধ অদ্ধক।
২১। আবহ মারুত ভিতরের বায় আগাং।
২২। আবহ পবনপবন অপরাজিত।
২৩। আবহ কাণ্ঠপ্রয় উধ্বগতি ধ্রুব।
২৪। আবহ নিশ্বাসক দ্বিগিল্দ্রয় ব্যাপ ব্যতিঘ।
২৫। আবহ উদান উদ্গৌরণ সক্ত।

২৬। পরিবহ অনিল অনুসাৃণ অশীত অজস্র।
২৭। পরিবহ সমীরণ পশ্চিমের বায় সরেষণ।
২৮। পরিবহ অনুষ্ণ শীতস্পর্শ পসন্দীক্ষ।
২৯। পরিবহ সন্ম্যাশ স্যথদা দেবদেব।
৩০। বিবহ বাতি বাক্ সম্ভব।
৩১। বিবহ অশ্রান্তি ধারণা অনিমিত্ত।
৩২। বিবহ প্রকল্পন কম্পন ভৌম।
৩৩। বিবহ সমান গোষণ একজ্যোতি।
৩৪। উতরহ মরুত উত্তর দিকের বায় সেনাজ্ঞ।
৩৫। উতরহ নভস্বান অপাকজ আভ্যন্ত।
৩৬। উহহ ধ্বনিঞ্জ আদিমিত।
৩৭। উতরহ কম্পলক্ষ্যা সেনাধন্বী।
৩৮। উতরহ বাস দেহব্যাপী বিধারণ।
৩৯। উতরহ মৃগবাহন বিদদ্বেবণ।
৪০। সংবহ চঞ্চল উৎকৃপঞ্চল দিব্যজ্যোতিঃাভ।
৪১। সংবহ অপান ক্ষুব্রাকর অধোগমন একশত।
৪২। সংবহ পৃষ্ঠাণ্ডগতি বর্গ মহাবল।
৪৩। বিবহ চপ্পর্শন চপ্পর্শ বিরাট।
৪৪। বিবহ বাত তিতষ্ণগমন পুত্রবানায়া।
৪৫। বিবহ প্রভজ্ঞান মনপৃথক সর্মিত।
৪৬। সংবহ অজগংপ্রাণ জন্মমরণ অদৃশ্য।
৪৭। সংবহ অবক ফেলা পূর্ণিমঃ।
৪৮। সংবহ সরীর প্রাতঃকালের বায় সর্মিত।
৪৯। সংবহ প্রকল্প গন্ধের অন্যেক যান মিতাসন।

দ্রষ্টব্য ঃ (ক) যে সব অক্ষর সহস্রায়ে রয়েছে সে সব অক্ষর ষট্‌চক্রে রেখাদ্বারা সংযুক্ত রয়েছে।
(খ) সহস্রারের তিন 'শ' মূলাধারের 'শ' এর সহিত মিলিত হয়েছে।
(গ) সহস্রারের লিখিত 'হংসন্বয়ং গুরোর্দ্বার' ছবিতে অস্পষ্ট।

Figure 13

The Sahasra chakra, or energy centers, of Tantric literature. The diagram is used to understand how forty-nine types of breath are associated with the six centers. The thousand petals' center at the head is described as the "door of the Guru"—utter tranquility through inhaling and exhaling. Visually compare the nested geometrical objects with the Hermetic figures of Giordano Bruno. (Drawing by Yogi Lahiri Mahasaya [1828–1895] communicated to Swami Satyeswarananda Giri to Gary Adamson to me.)

CHAPTER ONE
Magic Construction

We are in the position of a little child entering a huge library whose walls are covered to the ceiling with books in many different tongues. . . . The child does not understand the languages in which they are written. He notes a definite plan in the arrangement of books, a mysterious order which he does not comprehend, but only dimly suspects.

–Albert Einstein, *The Saturday Evening Post*

Categories

When you gaze at magic squares with their amazing properties and hidden symmetries, it's sometimes difficult to believe that there are easy-to-remember ways to construct many of them using simple rules. In fact, handbooks of "mental magic" often give these methods as "secret" ways to impress audiences.[1] The mentalist shows a large, empty square to the enthralled onlookers and, with a flourish and a cry of "*voilà!*", creates a magic square with astonishing ease. Imagine yourself being able to write down a large magic square in under a minute in front of an adoring audience. In this chapter, I'll introduce you to a few methods for constructing some of the traditional magic squares. If you are interested in pursuing the construction of some of the more elaborate magic squares, several books[2] in the For Further Reading section provide additional material and food for thought.

Various methods for constructing magic squares have evolved through the ages. When considering these methods, it is useful to categorize magic squares in three classes:

- Magic squares of *odd order* (squares where the order N is of the form $2m + 1$, where m may be any positive integer 1, 2, 3, etc.).
- Magic squares of a *doubly even order* (where the order N is of the form $4m$, such as 4, 8, 12, 16, 32, etc.). The order of a doubly even square can be divided by 2 and 4.
- Magic squares of a *singly even order* (where N is of the form $2(2m + 1)$, such as 2, 6, 10, 14, 18, 22, etc.). The order of a singly even square can be divided by 2 but not 4.

Generation methods for these kinds of squares vary in complexity, and for brevity I focus mostly on magic squares of odd order. Corresponding general methods for the construction of even-order magic squares are difficult to implement with pencil and paper, and magic squares of singly even orders are generally the most difficult of all to construct.

De la Loubère's Method

The smallest possible square of an odd order greater than 1 is the third-order square from the Introduction that has only one arrangement if we discount reflections and rotations:

4	9	2
3	5	7
8	1	6

Third-order magic square

In 1693, mathematician Simon de la Loubère suggested a method to create any odd-order magic square. He learned of it while ambassador to Siam from the court of Louis XIV. You can think of this as the "upward-right method" for reasons that will become clear. Start by placing a 1 in the central upper cell. In this example, we will construct a fifth-order magic square.

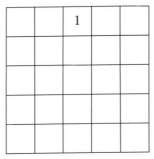

De la Loubère's first step

Proceed diagonally upward to the right and place a 2 in an imaginary square outside the table. To visually represent this process, let's temporarily add another row and column to the square. Because the 2 is outside the square, we bring it to the bottom of the column. Next, write a 3 upward to the right of the 2.

			2		
		1	↓		
	5		↓		
4	←	←	↓	←	4
			↓	3	
			2		

De la Loubère's movements

Next, write a 4 upward and to the right of the 3. Because 4 falls outside the original square, place it at the opposite end of the row, and write a 5 upward to the right. We cannot write a 6 upward and to the right of the 5 because the cell is occupied. Therefore, we write the 6 below the 5. We proceed until the 10 falls outside of the original square and continue the process. Notice that 16 falls outside the corner square and is written beneath the 15, as is the case when encountering an occupied square.

	18	25	2	9	
17	24	1	8	15	17
23	5	7	14	16	23
4	6	13	20	22	4
10	12	19	21	3	10
11	18	25	2	9	

De la Loubère's method completed

In summary

1. Start by placing 1 in the center cell, top row.
2. Whenever a top edge is reached, the integers are continued diagonally upward to the right at the bottom of the next column.
3. When the right-hand edge is reached, the integers are continued at the last cell to the left of the next highest row.
4. When a cell is reached that is already filled, drop down one cell and continue.
5. When the upper right corner cell is reached, drop down one row.

By rotations and reflections, seven other magic squares can be constructed from this method. Once you have memorized this simple approach, you can amaze your friends by generating odd-order squares of higher order. Additionally, you can start with a 1 in *any* cell and always generate a square that is magic in rows and columns but not necessarily diagonals.

You can also use de la Loubère's method to form imperfect magic squares that start with numbers other than 1. For example, the following square was created by starting with 3, and each succeeding integer was obtained by adding 2 to the preceding integer:

17	3	13
7	11	15
9	19	5

De la Loubère's square

Try this approach with other starting numbers and with other differences between succeeding integers.

It's also possible to use de la Loubère's method to form imperfect magic squares with more complicated number sequences. Let us define the horizontal difference ☞ as the difference between each succeeding integer in each group of *N* integers (represented as rows in the following squares). The vertical difference ♪ is defined as the difference between the last integer of the first *N* integers and the first integer of the next set of *N* integers. ☞ and ♪ do not have to be equal but must remain constant throughout. So you can more easily get a feel for these parameters, here are some examples of ☞ and ♪ in number arrays. (These are not magic squares.)

1	2	3	4
5	6	7	8
9	10	11	12
13	14	16	16

☞ = 1, ♪ = 1

1	2	3	4
6	7	8	9
11	12	13	14
16	17	18	19

☞ = 1, ♪ = 2

2	4	6	8
9	11	13	15
16	18	20	22
23	25	27	29

☞ = 2, ♪ 1

Here is a magic square created by de la Loubère's method using a starting number of 3, where ☞ = 2 and ♪ = 3:

19	3	14
7	12	17
10	21	5

Magic square, ☞ = 2, ♪ = 3

In other words, the groups 3, 5, 7; 10, 12, 14; and 17, 19, 21 were used to create this square.

When a magic square of order N is constructed in these ways, it is possible to predict in advance the magic constant S for the square using the formula

$$S = \tfrac{1}{2}(N^3 + N) + N(\square - 1) + (\kappa - N)[N(\varpi - 1) + (\phi - 1)]$$

where S is the magic constant, N is the order of the square, $\square$ is the number in the starting cell, κ is the sum of the numbers $1 + 2 + 3 + \cdots + N$, ϖ is the horizontal difference, and ϕ is the vertical difference. For the previous magic square, we get $S = \tfrac{1}{2}(3^3 + 3) + 3(3 - 1) + (6 - 3)[3(2 - 1) + (3 - 1)] = 15 + 6 + 3 \times (3 + 2) = 36$. Notice that for a pure magic square, $\square = \phi = \varpi = 1$, and the formula simplifies to the one provided in this book's introduction.

The formula becomes much simpler if you create a magic square by starting with a number other than 1 and filling all cells with consecutive numbers. If we denote the number in the starting cell by $\square$ and the order by N, the magic constant of this kind of square can be determined by

$$S = \tfrac{1}{2}(N^3 + N) + N(\square - 1)$$

Claude Gaspar Bachet de Méziriac's Method

Claude Gaspar Bachet de Méziriac (1581–1638) used a method similar to de la Loubère's to generate magic squares. Bachet joined the Jesuit Order in 1601 but left the order the following year after an illness. For most of his life Bachet lived in comparative leisure on his estate at Bourg-en-Bresse. He wrote books on mathematical puzzles and tricks that formed the basis of many later books on mathematical recreations.

To use Bachet's method, start with a 1 in the cell just above the center cell. Follow the diagonal rule of de la Loubère's method until an occupied cell is reached.

23	6	19	2	15
10	18	1	14	22
17	5	13	21	9
4	12	25	8	16
11	24	7	20	3

17	24	1	8	15
23	5	7	14	16
4	6	13	20	22
10	12	19	21	3
11	18	25	2	9

Bachet's method   De la Loubère's method

When this condition arises, as with the 5 in this example, the integers are continued in the same column *two cells higher.* If this is not possible because it would be above the magic square, as with 10, move down to the bottom cell in that column instead. When the upper right corner cell is reached, as with 15, the sequence is continued in the cell just above the right bottom corner cell. A comparison of Bachet's method and de la Loubère's method is shown here for a fifth-order square.

The following 15×15 square was constructed in the early 1600s.[3] It was reported in the *Arithmetischer Cubic-cossisher Lustgarten* ("Arithmetic-Algebraic Pleasure Garden," Tübingen, 1604) by the German mathematician and Rosicrucian Johann Faulhaber (1580–1635). (The Rosicrucians are a brotherhood combining elements of mystical beliefs with an optimism about the ability of science to improve the human condition.) The square's author is unknown, and I include it here because it provides an instructive example of a construction method for odd-order magic squares that is similar to Bachet's method.

Start with a 1 in the cell to the right of the center cell, and continue diagonally upward to the right. A diagonal that reaches the right-hand border column of the square is continued at the left-hand side of the square. Similarly, when a diagonal reaches the top row of the square, it is continued in the bottom row. These two rules are illustrated by the jump between 7 and 8 and by the jump between 8 and 9. When a diagonal encounters a cell that is already occupied, it restarts at the cell two steps to the right in the same row. This is shown by the jump between 15 and 16, where the 1 cell is occupied. An interesting case occurs at cell 120 at the upper

8	121	24	137	40	153	56	169	72	185	88	201	104	217	120
135	23	136	39	152	55	168	71	184	87	200	103	216	119	7
22	150	38	151	54	167	70	183	86	199	102	215	118	6	134
149	37	165	53	166	69	182	85	198	101	214	117	5	133	21
36	164	52	180	68	181	84	197	100	213	116	4	132	20	148
163	51	179	67	195	83	196	99	212	115	3	131	19	147	35
50	178	66	194	82	210	98	211	114	2	130	18	146	34	162
177	65	193	81	209	97	225	113	1	129	17	145	33	161	49
64	192	80	208	96	224	112	15	128	16	144	32	160	48	176
191	79	207	95	223	111	14	127	30	143	31	159	47	175	63
78	206	94	222	110	13	126	29	142	45	158	46	174	62	190
205	93	221	109	12	125	28	141	44	157	60	173	61	189	77
92	220	108	11	124	27	140	43	156	59	172	75	188	76	204
219	107	10	123	26	139	42	155	58	171	74	187	90	203	91
106	9	122	25	138	41	154	57	170	73	186	89	202	105	218

Johann Faulhaber's fifteenth-order magic square

right, where the progress of the diagonal is blocked by cell 106 at the lower left. Because 120 is the last cell in the top row, the two-step shift of the diagonal has to be made at the left end of the top row, and 121 is entered in the second cell from the left. The diagonal is then continued from cell 122 in the bottom row.

The procedure is continued until the last number, 255, is entered two steps away from the starting cell with 1. The sum of *all* the numbers in any fifteenth-order magic square is $N^2(N^2 + 1)/2 = 225 \times 226/2 = 25,425$. The sum for every row, column, and main diagonal is $25,425/15 = 1695$.

The eclectic Johann Faulhaber was fascinated by magic squares even though he was trained as a weaver. Later, he was taught mathematics in Ulm, Germany, and showed such promise that the city of Ulm appointed him city mathematician and surveyor. He opened his own school in Ulm and was in great demand because of his skill in fortification work. His expertise enabled him to work on fortifications in Basel, Frankfurt, and many other cities. He also designed waterwheels in Ulm and made mathematical and surveying instruments, particularly ones with military applications.

Faulhaber was an algebraist and important for his work explaining logarithms and sums of powers of integers. Faulhaber collaborated with many eminent people of his day, such as astronomer Johannes Kepler and Ludolph van Ceulen, the German mathematician who calculated pi to thirty-five places. His scientific and Rosicrucian beliefs made a major impression on philosopher René Descartes.

Philippe de la Hire's Method

French mathematician Philippe de la Hire (1640–1719) translated Emanuel Moschopoulos's essays on magic squares and collected many magic square theorems. De la Hire's method of creating magic squares of singly even order makes use of two generating squares that get summed together.

6	2	3	4	5	1
1	5	3	4	2	6
1	2	4	3	5	6
6	2	4	3	5	1
1	5	4	3	2	6
6	5	3	4	2	1

A

0	30	0	30	30	0
6	6	24	24	6	24
18	12	12	12	18	18
12	18	18	18	12	12
24	24	6	6	24	6
30	0	30	0	0	30

B

6	32	3	34	35	1
7	11	27	28	8	30
19	14	16	15	23	24
18	20	22	21	17	13
24	29	10	9	26	12
36	5	33	4	2	31

C

Let's try to use his method to construct a sixth-order magic square *C* from two other 6 × 6 squares *A* and *B*. In square *A*, fill the cells in the main diagonals (highlighted) with the numbers 1 through N beginning at the top of one diagonal and at the bottom of the other diagonal. In our sixth-order example, $N = 6$, and we fill the diagonals with numbers 1 through 6. In the next step, you have lots of leeway. Fill each of the remaining cells of the first column with either N or its "complement" (the integer 1) in any way you like, providing that there is the same number of each of these integers in the column. (Here we use the term "complement" of number x to mean $N - x + 1$, although authors also sometimes use "complement" when referring to $N^2 - x + 1$.) In our example, we fill in the first column with three 1s and three 6s.

Now that we have filled in column 1, it's time to fill in column N with the complements of column 1. This means wherever you see a 6 in column 1, place a 1 in column 6, and vice versa. This generates the rightmost column: 1, 6, 6, 1, 6, 1.

Fill the remaining columns on the left side of the center line using the numbers already in the column, or their complements, again with the caveat that there must be equal numbers of each integer in the column. Columns on the right side of the center are filled with the complements of the left side. (You can imagine a mirror plane coming out of the paper that reflects numbers on the left with their complements on the right.)

Now let's fill square B in a manner similar to the method for the first square except that only the integers $(0, N, 2N, 3N, \ldots, (N-1)N)$ are used. (Example: For a 3×3 square, we would use the integers 0, 3, and 6.) Begin by filling each diagonal, starting from the upper left and upper right with the numbers $(0, N, 2N, 3N, \ldots, (N-1)N)$. Fill the remaining cells of the first row with integers that are already in two of them or their complements. Continue as in square A, except this time a horizontal mirror reflects all of the numbers above the square's middle with their complements below.

Finally, add corresponding numbers in A and B to get magic square C.

Similarly, the de la Hire method can be used to construct doubly even order squares. The first square A uses the numbers 1 through N arranged in a way as we just discussed. The N integers used in the second square *B* are 0, *N*, 2*N*, 3*N*, ... , (*N* − 1)*N* as discussed. Add corresponding numbers in *A* and *B* to get magic square *C*.

1	3	2	4
4	2	3	1
4	2	3	1
1	3	2	4

A

0	12	12	0
8	4	4	8
4	8	8	4
12	0	0	12

B

1	15	14	4
12	6	7	9
8	10	11	5
13	3	2	16

C

We can construct Albrecht Dürer's square from the Introduction using a similar approach.

4	3	2	1
1	2	3	4
1	2	3	4
4	3	2	1

A

12	0	0	12
4	8	8	4
8	4	4	8
0	12	12	0

B

16	3	2	13
5	10	11	8
9	6	7	12
4	15	14	1

C (Dürer's square)

Various authors have used variants of de la Hire's method to construct magic squares of any order. For example, the first two squares that follow (*A* and *B*) are used to generate the magic square on the right (*C*):

3	1	4	2	5
5	3	1	4	2
2	5	3	1	4
4	2	5	3	1
1	4	2	5	3

A

15	0	20	5	10
0	20	5	10	15
20	5	10	15	0
5	10	15	0	20
10	15	0	20	5

B

18	1	24	7	15
5	23	6	14	17
22	10	13	16	4
9	12	20	3	21
11	19	2	25	8

C

Notice that in the leftmost array *A*, the numbers 1 through 5 are arranged so that every number appears once and only once in every row and column and one main diagonal. The other diagonal is composed of all 3s. Notice that all the short left diagonals repeat the same number. In array *B*, the numbers 0, 5, 10, 15, and 20 (0, $N, 2N, 3N, \ldots, (N-1)N$) are treated in the same way; in this case 10s are repeated along a main diagonal. Notice that all the right short diagonals repeat the same number. If these squares are superimposed and added, they produce the magic square on the right.

Stairstep Method

You may construct a magic square of odd order using the following easy recipe that involves a staircaselike assembly of cells. As an example, let's construct a 3×3 magic square. First draw a "staircase" with consecutive integers along diagonals like the following:

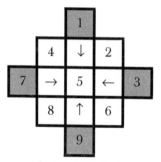

Stairstep method

The cells with 1, 3, 7, and 9 are outside the central 3×3 array. Slide the numbers in these outliers to the vacant cells furthest away along the same row or column. This produces the following magic square:

4	9	2
3	5	7
8	1	6

Third-order magic square

Here is another example of the staircase method, this time used to construct a 5×5 magic square:

		25		
	24		20	
23	•	19	•	15
22	•	18	•	14	•	10
21	17	•	13	•	9	5
16	•	12	•	8	•	4
11	•	7	•	3
	6		2
		1	

Staircase before "infolding"

23	6	19	2	15
10	18	1	14	22
17	5	13	21	9
4	12	25	8	16
11	24	7	20	3

Final 5×5 magic square

This time, let's start with the 1 on the bottom cell and proceed with numbering diagonally to the right. The dots represent empty cells to be filled in a manner similar to the previous 3×3 cell example.

John Lee Fults's Triangular-Parallelogram Method

John Lee Fults, author of *Magic Squares*, invented a method that can be used to construct odd-order magic squares. Write the integers 1 through

N^2 starting at the upper left as in the following array. Next chop off four "triangular" corners of the square, which leaves an internal square that is rotated so that its corner sticks up. (In this example, the 3 sticks up.)

1	2	3	4	5
6	7	8	9	10
11	12	13	14	15
16	17	18	19	20
21	22	23	24	25

Fult's method

For a 5×5 square, you chop off (L-shaped) triangles of three cells each. For a 7×7 square you chop off triangles of six cells each. In general, the corner triangles contain $(N^2 - 1)/8$ cells.

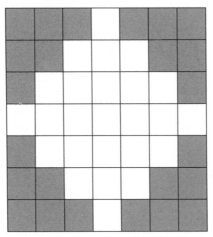

Cell-chopping in a seventh-order square

Next, you simply move each of the cut-off triangles along the main diagonal to opposites sides of the internal square. For example, in this case the upper left triangle composed of the numbers 1, 2, and 6 moves downward to the right to the position of the three dots. Notice how this creates the bottom right group (23, 6, 19, 2, 15)

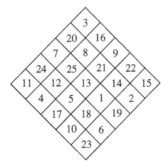

Figure 1.1
Diamond magic square.

in Figure 1.1. The triangle at the upper right composed of the numbers 4, 5, and 10 moves downward to the left. Move the other two triangles in a similar manner. When the moves are completed, you will get a diamond-shaped magic square (Fig. 1.1). Why not try this for other odd orders? If you attempt to construct a 3×3 square, the "triangles" you chop off will consist only of a single cell at each corner in the original square.

Ralph Strachey's Method

Here's a method that you can use to construct certain magic squares of singly even order.[4] As discussed, this class of squares has an order that can be divided by 2 but not divided by 4. For example, magic squares of order 6, 10, and 14 are singly even-order squares. Recall that singly even squares of order N are defined such that $N = 4x + 2$ for positive integer values of x. In the following example, $N = 6$ and $x = 1$. The Strachey method is not quite as elegant as for odd squares in the sense that the method is largely empirical. Singly even magic squares are a big challenge to construct using simple rules.

Let's describe the method generally before giving a specific example. Start by subdividing a square into four equal parts, A, B, C, and D.

A	C
D	B

Strachey's method

Eventually we will insert square arrays into A, B, C, and D; these four square arrays will be of odd order M each with $N^2/4$ cells. For our example case of $N = 6$, this means that each subsquare is of order $M = 3$, with nine cells.

Here's a general description. Start by using de la Loubère's method to construct a magic square with integers 1 through M^2 where $M = N/2$. Jam this magic square into A. Construct and jam three other magic squares into B, C, and D. These additional magic squares go from $M^2 + 1$ through $2M^2$, $2M^2 + 1$ through $3M^2$, and $3M^2 + 1$ through $4M^2$. You will now have a large square filled with consecutive integers 1 through N^2. If you examine this $N \times N$ square, you'll find that the sum R of the integers in each row is the same. The sum C of the integers in each column is the same, but R does not equal C. The sums D_1 and D_2 of the integers in the main diagonals are different from the column sums or row sums. By swapping a small number $(N^2/4 - N)$ of integers, you can convert the square into a magic square.

Sound complicated? Let's look at an example to clarify this mumbo-jumbo. I remind you that these singly even squares are all of order $4x + 2$. Let's choose $x = 1$, which gives us a sixth-order square. Start by dividing a square into four equal squares A, B, C, and D. Using de la Loubère's method, we use the numbers 1 through 9 to construct the upper left square in the following array:

8	1	6	26	19	24
3	5	7	21	23	25
4	9	2	22	27	20
35	28	33	17	10	15
30	32	34	12	14	16
31	36	29	13	18	11

Strachey's method

In square B, we use the same approach to construct a magic square using numbers 10 through 18. In square C, we use the same approach to construct a magic square using numbers 19 through 27. In square D, we use the same approach to construct a magic square using numbers 28 through 36. Look carefully. We seem to be close to having a 6×6 magic square. Notice that at this point, all rows sum to the same number, 84. All columns sum to 111. The right main diagonal sums to 165. The left main diagonal sums to 57. We can turn this 6×6 square into a magic square by swapping just three pairs of numbers. To help you see the swaps, I've singly and doubly underlined pairs and put a pair in bold italic. Just by swapping the 8 and 35, the 4 and 31, and the 5 and 32 we get a perfect sixth-order magic square. You can make a tenth-order magic square with $(N^2/4 - N) = 15$ swaps within four 5×5 subsquares, but because the exact pattern of swapping seems to be something arbitrary, that must be memorized, I won't go into further detail here. For additional examples, see Fults.[5] Perhaps you or some other farsighted person in the future will be able to discover a method for creating this kind of square in a more systematic manner.

The Diagonal Method

The diagonal method is a quick and easy way to make doubly even magic squares. All you have to do is write down the numbers 1 through N^2 in order, starting at the upper left cell. Highlight the two main diagonals and also the main diagonals of every 4×4 nonoverlapping subsquare that composes the main square. (More precisely, the nonoverlapping subsquares are formed when the $4k \times 4k$ array is divided into k^2 4×4 subsquares.) All numbers in cells along the diagonals are swapped with skew-related cells. (Recall from the Introduction that cells placed at equal distances on opposite ends of an imaginary line through the square's center are called skew-related cells.) That's all there is to it! An example will help clarify.

1	2	3	4
5	6	7	8
9	10	11	12
13	14	15	16

16	2	3	13
5	11	10	8
9	7	6	2
4	14	15	1

Before diagonal swaps   After diagonal swaps

In this example with a 4×4 square, we first write the consecutive numbers 1 through 16. Next we examine the main diagonals and make symmetrical swaps across the center of each highlighted diagonal. This means 1 swaps with 16, 6 with 11, 13 with 4, and 10 with 7. Isn't that a great way to make a magic square? Impress your friends by using this approach on a huge square, such as the following 12×12 example.

1	2	3	4	5	6	7	8	9	10	11	12
13	14	15	16	17	18	19	20	21	22	23	24
25	26	27	28	29	30	31	32	33	34	35	36
37	38	39	40	41	42	43	44	45	46	47	48
49	50	51	52	53	54	55	56	57	58	59	60
61	62	63	64	65	66	67	68	69	70	71	72
73	74	75	76	77	78	79	80	81	82	83	84
85	86	87	88	89	90	91	92	93	94	95	96
97	98	99	100	101	102	103	104	105	106	107	108
109	110	111	112	113	114	115	116	117	118	119	120
121	122	123	124	125	126	127	128	129	130	131	132
133	134	135	136	137	138	139	140	141	142	143	144

Before diagonal swaps

144	2	3	141	140	6	7	137	136	10	11	133
13	131	130	16	17	127	126	20	21	123	122	24
25	119	118	28	29	115	114	32	33	111	110	36
108	38	39	105	104	42	43	101	100	46	47	97
96	50	51	93	92	54	55	89	88	58	59	85
61	83	82	64	65	79	78	68	69	75	74	72
73	71	70	76	77	67	66	80	81	63	62	84
60	86	87	57	56	90	91	53	52	94	95	49
48	98	99	45	44	102	103	41	40	106	107	37
109	35	34	112	113	31	30	116	117	27	26	120
121	23	22	124	125	19	18	128	129	15	14	132
12	134	135	9	8	138	139	5	4	142	143	1

After diagonal swaps

This is a little bit trickier because in addition to swapping cells along the main diagonals, we also have to locate cells that are in the diagonals of all the 4×4 subsquares, one of which is shown here at upper right, and swap their values with skew-related cells. For example, 4 and 141 are swapped, as well as all other cells not already swapped in the main diagonals, such as 37 and 108, 26 and 119, and 15 and 130. When we're finished, we generate the magic square on the right with the magic constant $\mathcal{S} = N(N^2 + 1)/2 = 870$.

Dürer's Method

In the Introduction we discussed Albrecht Dürer's 4×4 magic square constructed in 1514 A.D. At that time Dürer described a general method for constructing doubly even magic squares. First, draw imaginary main diagonals through every 4×4 subsquare of the square such as in the following square on the left. (This is similar to the diagonals drawn in the previous section.) Place the number 1 in the upper left corner cell and proceed horizontally to the right with consecutive numbers, but only write the numbers down when the cells are crossed by diagonals. Continue this procedure for each row. Once you arrive at the bottom right corner cell, pretend that the number 1 falls on this square and reverse the process, proceeding horizontally to the left and filling the unfilled cells. For a fourth-order square we have the following before-and-after diagrams:

1			4
	6	7	
	10	11	
13			16

1	15	14	4
12	6	7	9
8	10	11	5
13	3	2	16

Durer's method, before   Durer's method, after

Knight's Move Method

One elegant method[6] for constructing a variety of odd-order squares involves the knight's move in chess (which is a move two cells horizontally or vertically and then one cell at a right angle). In our application, the knight's move is constrained to two cells upward and one cell to the right. Start by placing a 1 in the center cell of the upper row. Pretend that the magic square is wrapped around a doughnut so that the top cells meet the bottom cells and right cells meet left cells, and start galloping your horse.

We'll use a fifth-order magic square for our example. The horse gallops off the square at the top and wraps around to the cell labeled 2 at the bottom. Next, the horse hops to the 3 and then 4 and 5. The next move is blocked by 1, so 6 is written below 5. The horse continues until the array is filled.

10	18	1	14	22
11	24	7	20	3
17	5	13	21	9
23	6	19	2	15
4	12	25	8	16

Knight's move method

In this square, each row, column, and main diagonal sums to 65. Also note that the sum of the two numbers in any two skew-related cells is always equal to a constant sum, 26, which is twice the value of the center square or half the magic constant. Again, as a reminder, the skew-related cells are on opposite ends of an imaginary line drawn through the center and include such pairs as (6, 20) and (23, 3). Try varying this method. For example, you should be able to produce magic squares in which the knight's move may be upward and to the left instead of to the right, or downward to the right or left.

Lozenge Method

Another method for generating odd-order magic squares was discussed by mathematician John H. Conway under the name "lozenge" method,[7] and the method appears to date back at least to the early 1900s.[8] To use this approach, create an interior diamond filled with odd numbers as in the following diagram. Start with 1 at the leftmost vertex and go diagonally upward to the right. Continue filling diagonally until the diamond is completed as shown with the bold numbers.

		2	8	
18	24	5	6	12
22	3	9	15	16
1	7	13	19	25
10	11	17	23	4
14	20	21	2	8

Lozenge method

The even numbers that were missed are now added sequentially along the continuation of the diagonals, with certain wrapping rules applied when the edge of the magic square is reached. For example, consider the first diagonal 1, 3, 5. We are missing the numbers 2 and 4. To determine where these go, continue the diagonal upward and to the right. This takes us to an imaginary square above the 6. Whenever you find yourself in an imaginary square above a column, slide down. (When you are in an imaginary square to the left of a row, slide to the left.) This means the 2 slides down to the bottom row. The diagonal is continued to the right for the 4. If we continue diagonally to the right, this brings us back to our starting 1 using the wrap rule. This means we must start again with the next diagonal, which contains 7 and 9. Continue up to the right with the missing number 6. Continue off the board with an 8, which wraps to the bottom. Continue up to the right and place a 10 outside the square, which wraps to the left. Continue up and to the right. Because we have reached the starting point, we consider the next diagonal of odd numbers. The process continues until the square is filled with the missing even numbers.

The Diagonal Rule for Odd-Order Magic Cubes

Although I focus on magic squares in this chapter, I'd like to whet your appetite for constructing higher-dimensional figures such as magic cubes. A magic cube is a cube composed of N^3 consecutive numbers, where N is the order of one of the squares on the cube's faces. The numbers used in forming a magic cube are usually 1

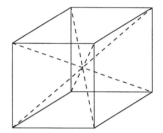

Figure 1.2
The four triagonals, or space diagonals, of a cube.

through N^3. They are arranged so that each row, each column, each main diagonal of the square cross sections, and each of the four great (space) diagonals (sometimes called "triagonals") containing N integers will add to the same sum (Fig. 1.2). (Magic cubes are thoroughly defined in chapter 2.) In 1988, John Hendricks developed a simple approach for constructing odd-order magic cubes.[9]

To create a magic cube, start by placing the number 1 at the center of the bottom layer of the cube (see Fig. 1.3). Next, begin to fill the vertical center plane as shown in Figure 1.2. You fill in the numbers on the plane just as in a magic square, proceeding diagonally downward to the right. When you run off the bottom of the plane, you reenter at the top. When you run off the right-hand side, you reenter on the left-hand side.

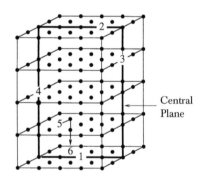

Figure 1.3
Creating a magic cube using the diagonal rule. In the cubic array, place 1 in the center of the bottom layer, then proceed diagonally down to the right along a central vertical square cross section, as in the diagonal rule for squares, to place 2, 3, 4, and 5. When you can no longer proceed, go back one layer and down one to place the number 6. This backward move is represented by the arrow from 5 to 6. Then continue diagonally and repeat the jump step as required.

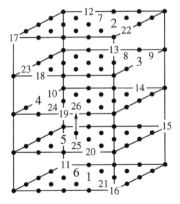

Figure 1.4

Additional steps in cube creation. You have proceeded until reaching a double impasse that occurs at 25. Move up one cell every time you have completed N^2 numbers, as indicated by the arrow. Continue moving diagonally until the cube is complete, using the "back one, down one" routine for a single impasse and the "up one" routine for a double impasse.

Each time there is a number in the way, move back one layer and down one, as shown in Figure 1.3 for the move between 5 and 6. In other words, because 1 blocks the 6, the 6 is shifted backward and down as indicated by the arrow. This motion is called a jump-step. Continue to proceed diagonally in this new plane, repeating the jump-step, one back and one down, as required. Eventually, after N planes and N^2 numbers, another impasse is reached. At this point go up one, as indicated by the arrow between 25 and 26 in Figure 1.4.

Use the "up one" procedure only if you cannot use the "back one and down one" procedure, that is, at the point where the number is divisible by N^2. Eventually, you will arrive at a fifth-order cube of magic sum 315 with the following layers:

110	12	44	71	78
98	105	7	39	66
61	93	125	2	34
29	56	88	120	22
17	49	51	83	115

Top layer of cube

79	106	13	45	72
67	99	101	8	40
35	62	94	121	3
23	30	57	89	116
111	18	50	52	84

Second layer of cube

73	80	107	14	41
36	68	100	102	9
4	31	63	95	122
117	24	26	58	90
85	112	19	46	53

Third layer of cube

42	74	76	108	15
10	37	69	96	103
123	5	32	64	91
86	118	25	27	59
54	81	113	20	47

Fourth layer of cube

11	43	75	77	109
104	6	38	70	97
92	124	1	33	65
60	87	119	21	28
48	55	82	114	16

Bottom layer of cube

Using Hendricks's approach, you can write out magic cubes of odd order as large as you want! Now that's something to impress your buddies with at the next stockholders' meeting.

Other Methods

Many other methods for constructing magic squares exist, including Arnoux's method (discovered in 1894) and Margossian's method (discovered in 1908).[10] These novel approaches have been discussed in the literature, for example in *Magic Squares* by John Lee Fults, *Magic Squares and Cubes* by W. S. Andrews, and W. W. Ball and H. S. M. Coxeter's *Mathematical Recreations and Essays*. Most writers develop their favorite methods, but the holy grail of magic squares creation would be to discover a method that would generate every possible arrangement for a square of a given size. Such a solution is probably not discoverable.

One of my favorite, little-known ways of generating one magic square from another involves matrix multiplication. In 1992, Frank E. Hruska, a Professor of Chemistry at the University of Manitoba,[11] suggested that if a third-order magic square is treated as a matrix, then cubing the matrix will produce another matrix square. The following is an example:

4	9	2
3	5	7
8	1	6

3 × 3 matrix, $S = 15$

1149	1173	1053
1029	1125	1221
1197	1077	1101

3 × 3 matrix cubed, $S = 3375$

On the left is a magic square that produces the magic square on the right when it is treated as a matrix and cubed. Recall that for matrix multiplication you don't simply cube each number in the matrix. The product C of two matrices A and B is defined by

$$c_{ij} = a_{i1}b_{1j} + a_{i2}b_{2j} + \cdots + a_{in}b_{nj} = \sum_{k=1}^{n} a_{ik}b_{kj},$$

where j is summed over for all possible values of i and k. Here is an example to refresh yourself about matrix multiplication.

$$\begin{bmatrix} 3 & -2 & 1 \\ 2 & -1 & 3 \end{bmatrix} \times \begin{bmatrix} -1 & 2 \\ 4 & -1 \\ 5 & 6 \end{bmatrix} = \begin{bmatrix} 3(-1)+(-2)4+1(5) & 3(2)+(-2)(-1)+1(6) \\ 2(-1)+(-1)4+3(5) & 2(2)+(-1)(-1)+3(6) \end{bmatrix} = \begin{bmatrix} -6 & 14 \\ 9 & 23 \end{bmatrix}$$

The previous magic square example shows that the matrix cubed generates a magic square with magic constant 3375. If you cube the matrix again, then you arrive at a magic square with magic sum 38,443,359,375. Isn't this a great, compact way to generate magic squares with large values?

12,814,121,349	12,815,780,229	12,813,457,797
12,813,789,573	12,814,453,125	12,815,116,677
12,815,448,453	12,813,126,021	12,814,784,901

Matrix cubed again, $\mathcal{S} = 38,443,359,375$

Now, if we cube this last matrix, we get the following behemoth magic square:

Magic Square Cell Values	Row	Column
1893837622053263813460921 7310949	1	1
1893837622052825569291689 0521829	1	2
1893837622053439111128614 8026597	1	3
1893837622053351462294768 2668773	2	1
1893837622053176164627075 1953125	2	2
1893837622053000866959382 1237477	2	3
1893837622052913218125535 5879653	3	1
1893837622053526759962461 3384421	3	2
1893837622053088515793228 6595301	3	3

Matrix cubed again

The magic constant is

56,815,128,661,595,284,938,812,255,859,375

Colleague Phil Hobbs and I notice that raising any third-order magic square to any odd power seems to yield a magic square. (We used a computer to test this up to the fifteenth power.) However, simply squaring the matrix does not produce a magic square. Interestingly, all the even powers have a peculiar structure, namely, all the elements on the main diagonal are the same, and all the off-diagonal elements are the same, too. Thus, the even powers fail to be a magic square because the main diagonal sums are different from the other sums, and because the elements are not all different numbers.

It appears that you can easily demonstrate that cubing any 3×3 magic square (considered as a matrix) produces another magic square.[12] Consider that any 3×3 magic square can be represented in one of two forms. For example, one form by John Hendricks can be represented as

$y - a$	$y + a + b$	$y - b$
$y + a + b$	y	$y - a + b$
$y + b$	$y - a - b$	$y + a$

Order-3 matrix

Here the magic sum is $3y$. In the original example 3×3 square, $y = 5$, $a = 1$, and $b = 3$. The cube of this square is

$9y^3 - 3a^3 + 3ab^2$	$9y^3 + 3(a^2 - b^2)(a + b)$	$9y^3 - 3a^2b + 3b^3$
$9y^3 + 3(a^2 - b^2)(a - b)$	$9y^3$	$9y^3 - 3(a^2 - b^2)(a - b)$
$9y^3 + 3a^2b - 3b^3$	$9y^3 - 3(a^2 - b^2)(a + b)$	$9y^3 + 3a^3 - 3ab^2$

Matrix cubed

This result is also always a magic square with magic sum $27y^3$.

Allan Johnson, Jr., notes that you can also write 3×3 magic squares in another form:[13]

$p + s + t$	$p - 2s$	$p + s - t$
$p - 2t$	p	$p + 2t$
$p - s + t$	$p + 2s$	$p - s - t$

Order-3 matrix

When this square is treated as a matrix and cubed, we get

$9p^3 + 12s^2t + 12st^2$	$9p^3 - 24s^2t$	$9p^3 + 12s^2t - 12st^2$
$9p^3 - 24st^2$	$9p^3$	$9p^3 + 24st^2$
$9p^3 - 12s^2t + 12st^2$	$9p^3 + 24s^2t$	$9p^3 - 12s^2t - 12st^2$

Matrix cubed

This result is also always a magic square with magic sum $27p^3$.

I believe the field is wide open for study with respect to cubing higher-order magic squares. Consider the following example experiment in which a 4 × 4 pandiagonal magic square (S = 34) yields another magic square (S = 39,304) when cubed.[14] Again, the squared matrix is not a magic square. One wonders which magic squares of arbitrary size yield magic squares when treated as a matrix and cubed. Experiments show that not all 4 × 4 magic squares turn into magic squares when cubed.

1	15	4	14
12	6	9	7
13	3	16	2
8	10	5	11

4 × 4 matrix, S = 34

9226	10,106	10,186	9786
10,346	9626	9386	9946
9466	9866	10,426	9546
10,266	9706	9306	10,026

4 × 4 matrix cubed, S = 39,304

CHAPTER TWO
Classification

Magic squares are conspicuous instances of the intrinsic harmony of number, and so they will serve as an interpreter of the cosmic order that dominates all existence. Though they are a mere intellectual play, they not only illustrate the nature of mathematics but also the nature of existence dominated by mathematical regularity.

–Paul Carus, in W. S. Andrews's *Magic Squares and Cubes*

Magic squares can be classified in many different ways according to special properties they may possess. For example, squares are often placed in four major categories: Nasik, associated, simple, and semi-Nasik. As will become evident, some magic squares may have more than one classification. Several of the weird, modern magic squares in chapter 3 don't easily fall into any of the broad classes of magic squares that have interested mathematicians for decades, and in some cases for centuries.

Simple and Semimagic Squares

The *simple magic square* meets the minimum requirement that the sum of the integers in each row, column, and main diagonal is a constant. In a sense, this is the least magical of magic squares, and in the next few sections you'll see squares with additional properties that make them more magical. The *semimagic square* is a magic square *without* the main diagonal conditions.

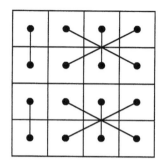

Figure 2.1
Pair connection diagram for a
simple magic square.

Here is an example simple magic square:

16	3	10	5
1	12	7	14
8	13	2	11
9	6	15	4

Simple magic square

Even with simple magic squares, there are also certain number pairs that sum to half the magic sum. The pairs are indicated as connected dots in Figure 2.1. These "pair connection diagrams" vary with the particular simple magic square being represented.

Of the 880 fourth-order magic squares, only 448 are simple, not counting rotations and reflections. It is not possible to construct a simple magic square (also called a *minimum requirement magic square*) for all orders. For example, there are only eight arrangements of a third-order square, and none of the eight are simple because they are classified as associate magic squares, defined in the following section.

Associated Squares

In addition to being a simple magic square, an *associated* (or *associative*) magic square also has skew properties. In particular, the sum of the two numbers in any two skew-related cells are always equal to a constant sum $N^2 + 1$. As already noted, skew-related cells are those cells at two ends of an imaginary line through the center of the square. In the square below, the skew-related cells sum to 17, for example, $1 + 16$, $2 + 15$, $3 + 14$, etc.

1	14	12	7
8	11	13	2
15	4	6	9
10	5	3	16

Associated square

Odd-order squares produced by the knight's move method discussed in the last chapter are associated, as is Albrecht Dürer's square from the Introduction. Figure 2.2 shows a pair connection diagram for associated squares. The number pairs sum to half the magic sum.

There are forty-eight associated fourth-order squares. All associated squares of the fourth order are also semi-Nasik, as you'll see in a following section.

To construct an associated square of odd order, most any method may be used that is used in the construction of odd-order

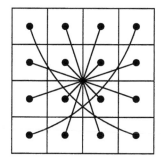

Figure 2.2
Pair connection diagram for an associated magic square.

magic squares. For example, the methods of de la Loubère, de la Hire, and others may be used to construct associated squares of odd order. You can use Dürer's method or other similar methods to construct doubly even-order associated squares.

Nasik (Diabolique or Pandiagonal) Squares

The following *Nasik magic square* was so named by Reverend A. H. Frost after the town in India where he lived and did missionary work. (Nasik is located in the northwestern Maharashtra state, western India, along the beautiful Godavari River; I'd be interested in hearing from readers who have visited Nasik.) Nasik magic squares are also known as *diabolique, diabolic, perfect,* or *pandiagonal* squares.

1	14	7	12
15	4	9	6
10	5	16	3
8	11	2	13

Nasik magic square

As with the simple and associated magic squares, certain number pairs sum to half the magic sum. The pairs are indicated as connected dots in Figure 2.3.

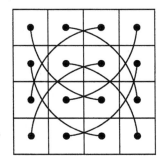

Figure 2.3
Pair connection diagram for a Nasik magic square.

I think that the Nasik square is the most elegant and "perfect" of all fourth-order squares because, in addition to the usual magic sums, all the *broken diagonals* sum to a constant. Broken diagonals leave one end of the square and return along another. An example broken diagonal is formed by the numbers 15, 14, 2, and 3. Notice that these broken diagonals all sum to 34, for example, $15 + 14 + 2 + 3 = 10 + 4 + 7 + 13 = 15 + 5 + 2 + 12 = 34$. But wait, the fun has just begun! If you repeat the square in all directions, you can then draw a box around any 4×4 array of numbers and it will be magic. Try it on the following array, where I've made a 16×16 array of numbers from four adjacent fourth-order Nasik squares. One 4×4 array is highlighted. Cut out an appropriately sized hole in a piece of paper and move it around on the 16×16 array. No matter where you put your frame, it will reveal a magic square.

1	14	7	12	1	14	7	12
15	4	9	6	15	4	9	6
10	5	16	3	10	5	16	3
8	11	2	13	8	11	2	13
1	14	7	12	1	14	7	12
15	4	9	6	15	4	9	6
10	5	16	3	10	5	16	3
8	11	2	13	8	11	2	13

Nasik array

Isn't that a dandy pattern suitable for wallpapering your room?

A Nasik square also remains Nasik under four different transformations: rotation, reflection, a transfer of a row from top to bottom or from bottom to top, or a transfer of a column from one side to the other. In the fourth-order Nasik square, we also find

1. Every 2×2 square drawn on an endless array of side-by-side squares adds up to 34.
2. Along every diagonal, any two cells separated by one cell add up to 17.

Let's examine a fifth-order Nasik square. Notice that the sum of any two integers occupying skew-related cells is 26. All broken diagonals add to 65, as well as all rows, columns, and main diagonals.

10	18	1	14	22
11	24	7	20	3
17	5	13	21	9
23	6	19	2	15
4	12	25	8	16

Fifth-order Nasik square

This fifth-order square has many marvelous properties. For example, the sum of the squares of the integers in the first and second rows equals the sum of the squares of the integers in the second and fourth rows, which equals the sum of the squares of the integers in the first and third rows, which equals the sum of the squares of the integers in the third and fourth rows, which equals the sum of the squares of the integers in the two main diagonals, which equals 2260! The same characteristic applies to the columns.

Following is an example of a fifth-order square frame that can be dropped anywhere on this infinite plane of numbers to form a magic square. I've highlighted just four Nasik magic squares, but you can create many more by moving the 5×5 frame around. (If the square is to be associative, it must contain a 13 at its center.)

1	15	24	8	17	1	15	24	8	17	1	15	24	8	17
23	7	16	5	14	23	7	16	5	14	23	7	16	5	14
20	4	13	22	6	20	4	13	22	6	20	4	13	22	6
12	21	10	19	3	12	21	10	19	3	12	21	10	19	3
9	18	2	11	25	9	18	2	11	25	9	18	2	11	25
1	15	24	8	17	1	15	24	8	17	1	15	24	8	17
23	7	16	5	14	23	7	16	5	14	23	7	16	5	14
20	4	13	22	6	20	4	13	22	6	20	4	13	22	6
12	21	10	19	3	12	21	10	19	3	12	21	10	19	3
9	18	2	11	25	9	18	2	11	25	9	18	2	11	25

Nasik array containing various Nasik magic squares

In general, a magic square is called Nasik, pandiagonal, or diabolic if all its broken diagonals add up to the magic constant. Such squares can be constructed of any odd order above 3 and of any order that is a multiple of 4 (that is of any doubly even order). However, a Nasik square of a singly even order has never been found and is considered impossible. There are forty-eight fourth-order Nasik squares, and all have similar pair connection diagrams. Excluding rotations and reflections, there are 3600 Nasik fifth-order magic squares.[1] If we also exclude variations obtained by cyclic permutation of rows and columns, 144 of these squares are pandiagonal. Of the 144, just 16 contain a square that is also associative. There are 38,102,400 seventh-order Nasik magic squares.[2]

The Lo-shu in the Introduction is associative but not pandiagonal. An order-4 square may be pandiagonal or associative but not both. The order-5 square is the smallest one that can have both properties. Of the sixteen associative pandiagonal squares of order 5, four have 1 in the first cell, four have 1 in the third cell, four have 1 in the seventh, and four have 1 in the eighth.[3] Moslems of the Middle Ages were fascinated by pandiagonal squares with 1 in the center. The patterns were not associative, but the Moslems thought of the central 1 as being symbolic of the unity of Allah. Indeed, they were so awed by that symbol that they often left blank the central cell on which the 1 should be positioned.

If a Nasik, or pandiagonal, square also has all the special properties of the order-4 pandiagonals, it is called *most perfect.* Here is a most-perfect eighth-order square:[4]

0	62	2	60	11	53	9	55
15	49	13	51	4	58	6	56
16	46	18	44	27	37	25	39
31	33	29	35	20	42	22	40
52	10	54	8	63	1	61	3
59	5	57	7	48	14	50	12
36	26	38	24	47	17	45	19
43	21	41	23	32	30	34	28

A most-perfect eighth-order square

The square is most perfect because it has a magic constant of 252, all its 2×2 subsquares add up to 126, and any two numbers that are $N/2 = 4$ cells apart add up to $N^2 - 1 = 63$. (In this square, I've started the square with 0 instead of 1. You'll find in the literature that magic squares are made with consecutive integers starting with 1 or 0. If it starts with 0, it can be changed to a square starting with 1 simply by adding 1 to each cell.)

Let's try to put most-perfect magic squares into perspective. Recall that all magic squares of order 3 are essentially the same because a rotation or a reflection of the square remains magic. There are many different magic squares of order N, and the number explodes as the order increases. No exact formula is known. Also recall that a magic square cannot be pandiagonal unless its broken diagonals also sum to a magic constant. Even-order magic squares cannot be pandiagonal unless their order is doubly even—that is, a multiple of 4. A *most-perfect* magic square is even more restricted. Not only is it magic and pandiagonal, it also has the property that any 2×2 block of adjacent entries sum to the same total, $2N^2 - 2$. (For convenience in this discussion, the magic squares of order N contain the integers 0, 1, 2, ..., $N^2 - 1$.) Note also that any magic square with this 2×2 summation property is necessarily pandiagonal.

Although all order-4 pandiagonals have been known to be most perfect for three centuries, little was known about most-perfect squares of higher order.[5] There was no method of constructing them all, or even of determining the number of squares of a given order. In 1998, magic square experts Dame Kathleen Olleren-shaw and David Brée finally settled these questions in their book *Most-Perfect Pandiagonal Magic Squares: Their Construction and Enumeration.*[6] In the book, they devise a method for constructing all most-perfect squares of any order and a way of calculating their number. Unlike the ordinary pandiagonals, there are no most-perfect squares with odd order so the only possible orders are multiples of four. As the order increases, the number of essentially different most-perfect squares skyrockets:

Order	Number of Most-Perfect Squares
4	48
8	368,640
12	2.22953×10^{10}
16	9.32243×10^{14}
36	2.76754×10^{44}

As Martin Gardner points out,[7] this last number is around a thousand times the number of pico-picoseconds since the Big Bang.

As I discuss in my book *Wonders of Numbers,* many useful mathematical discoveries are made by amateurs in a particular mathematical field. For the case of counting most-perfect magic squares, neither discoverer is a typical mathematician. Ollerenshaw spent much of her professional life as a high-level administrator for several English universities. Brée has held university positions in business studies, psychology, and artificial intelligence. Martin Gardner writes about Ollerenshaw and Brée's construction and enumeration methods for most-perfect pandiagonal magic squares:

> This solution of one of the most frustrating problems in magic-square theory is an achievement that would have been remarkable for a mathematician of any age. In Dame Kathleen's case it is even more remarkable, because she was 85 when she and Brée finally proved the conjectures she had earlier made. In her own words, "The manner in which each successive application of the properties of binomial coefficients that characterize the Pascal triangle led to the solution will always remain one of the most magical mathematical revelations that I have been fortunate enough to experience. That this should have been afforded to someone who had, with a few exceptions, been out of mathematics research for over 40 years will, I hope, encourage others. The delight of discovery is not a privilege reserved solely for the young."

Ian Stewart, Professor of Mathematics at the University of Warwick in England, called attention to Ollerenshaw and Brée's work noting that magic squares are still fertile ground for research:[8]

> It is astonishing what modern methods can achieve in such a traditional area. Even more surprising is the completeness of the results. This is a fascinating and totally unexpected accomplishment, eminently accessible to non-specialists. If you thought magic squares were mined out long ago, then think again.

We can get a glimmer of how Ollerenshaw and Brée tackled their enumeration of most-perfect magic squares by first considering the notion of a *reversible square*. A reversible square of order N is an $N \times N$ array containing integers 0 through $N^2 - 1$ with the following properties: (1) every row and column has *reverse similarity*, and (2) in any rectangular array of integers from the square, the sums of entries in opposite corners are equal. For example, the following 4×4 array is reversible:

0	1	2	3
4	5	6	7
8	9	10	11
12	13	14	15

A reversible square

An example of reverse similarity in condition (1) is $4 + 7 = 5 + 6 = 11$. The same relationship holds for all other rows and columns. Condition (2) is satisfied by such relations as $1 + 7 = 3 + 5 = 8$, as shown in the shaded rectangle. (Experiment with other rectangles to satisfy yourself that the corners have similar properties.) This reversible square is obviously not a magic square, but Ollerenshaw and Brée have proved that every reversible square of doubly even order can be changed to a most-perfect magic square by a specific procedure, and that every most-perfect magic square can be produced by these procedures.

To give you an idea about procedures that convert reversible squares to most-perfect magic squares, examine the following square on the left. To produce this square, we start with the previous 4×4 square and then reverse the (shaded) right half of each row:

0	1	3	2
4	5	7	6
8	9	11	10
12	13	15	14

First step

0	1	3	2
4	5	7	6
12	13	15	14
8	9	11	10

Second step

Next, we reverse the bottom half of each column (right square). Now, the tricky part. We break up the square into 2×2 blocks and move the four entries in each block as shown on the left in the following:

☺	💣		
👽	👽	👽	
	↓		💣
	👽		

0	1	3	2
4	5	7	6
12	13	15	14
8	9	11	10

Tricky procedure   At last—a most-perfect result

The meanings of the funny symbols are not too difficult to understand. The top left entry ☺ stays fixed. The top right entry 💣 moves diagonally two squares to the other 💣. The bottom left 👽 moves two squares to the right, and the bottom right 👽 moves two spaces down. This last downward movement is designated by ↓. If a number happens to fall off the edge of the 4×4 square, you wrap the edges around the square to find where it should go. Your result (the square on the right) is a most-perfect magic square. The method I just outlined, given in detail by Ian Stewart in the November 1999 issue of *Scientific American*, works only for order-4 squares. The general case of order N requires similar recipes expressed by a mathematical formula.

Stop and think for a moment. What have we achieved by this mathematical recipe? For one thing, this kind of transformation process establishes a one-to-one relationship between most-perfect magic squares and reversible squares of doubly even order. This means that you can count the number of most-perfect magic squares by counting the number of reversible squares of the same order. "Big deal," you say? It *is* a big deal because reversible squares have several features that make it possible to count them *all*. These features are outlined by Ollerenshaw and Brée, and this means, for the first time in human history, we finally have a way to count the number of most-perfect magic squares for any order.

Semi-Nasik (Semidiabolique) Squares

In *semi-Nasik squares*, also called *semidiabolique* or *semidiabolic squares*, the opposite "short diagonals" of two cells sum to a constant. In the example here, one pair of short diagonals is highlighted $(14 + 4 + 11 + 5 = 34)$. The other pair also sums to 34: $12 + 6 + 13 + 3 = 34$. In an odd-order square, the two opposite short diagonals contain $(N-1)$ cells, where N is the order of the square. In an even-order square, the two opposite short diagonals contain N cells. The sum of these N cells will be the same as the square's constant. The other broken diagonals of a semi-Nasik square, such as $13 + 15 + 12 + 10 = 50$, do not produce the magic sum.

1	14	12	7
4	15	9	6
13	2	8	11
16	3	5	10

Semi-Nasik square

Figure 2.4 is a pair connection diagram showing the number pairs that sum to half the magic sum for this particular semi-Nasik square. (The pair connection diagrams vary for different semi-Nasik

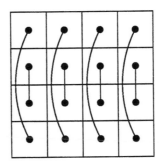

Figure 2.4
Pair connection diagram for a semi-Nasik magic square.

squares.) There are 384 fourth-order semi-Nasik squares (which include 48 associated and 448 simple), making a total of 880 fourth-order magic squares, if we do not consider reflections and rotations.

You can construct a semidiabolic square of odd order using de la Loubère's and de la Hire's methods, among others. If you use de la Loubère's method, the numbers in the two opposite short diagonals segments of $(N-1)/2$ cells plus the number occupying the center cell will sum to the square's constant. You can construct a semidiabolic square of doubly even order using Dürer's method, de la Hire's method, the diagonal method, and other, similar methods used to construct doubly even-order squares. All associated squares of the fourth order are semidiabolic; however all semidiabolic squares of the fourth order are not associated.

Here is a semidiabolic square constructed by de la Loubère's method:

17	24	1	8	15
23	5	7	14	16
4	6	13	20	22
10	12	19	21	3
11	18	25	2	9

Semidiabolic magic square

Notice that the opposite short diagonals have the same sums: $23 + 24 + 2 + 3 = 8 + 16 + 10 + 18 = 52$. If there were five numbers, then this would make a broken diagonal. When the center cell, 13, is added to the integers of the opposite short diagonals, the sum is the magic constant, 65.

Next let's examine the opposite short diagonals with six integers. When they are added together, their sum is $(4 + 5 + 1 + 25 + 21 + 22) + (1 + 14 + 22 + 4 + 12 + 25) = 78$. If 13 is subtracted from 78, we get the magic constant.

Bordered Squares

A *bordered* (or *concentric*) *magic square* contains all the usual proper-ties of a simple magic square plus the additional property that if one or more number "borders" are removed from the square a magic square still remains. As an example, consider this magic square consisting of a third-order magic square surrounded by a border so as to also form a square of fifth order.

2	23	25	7	8
4	16	9	14	22
21	11	13	15	5
20	12	17	10	6
18	3	1	19	24

Fifth-order bordered square

How rare do you think these wonders are? Perhaps not as rare as you would think. It turns out that there are 174,240 bordered squares of order 5 and 567,705,600 bordered squares of order 6. However, again we might want to think harder about what "rare" might mean for an array of numbers. For example, there are

$$1,938,901,255,416,373,248,000,000$$

and

$$46,499,165,848,737,652,183,499,931,018,854,400,000,000$$

possible 5×5 and 6×6 different arrays of numbers where no integers are used more than once and rotations and reflections are not considered. (As I mentioned in the Introduction, to compute these large numbers for any order N uses $N^2! / 8$, where "!" is the mathematical symbol for factorial.)

Here's one of my favorites, invented by French mathematician Bernard Frénicle de Bessy (1602–1675). In the Frénicle square, the borders may be successively peeled off to produce magic squares of the ninth, seventh, fifth, and third orders.

16	81	79	78	77	13	12	11	2
76	28	65	62	61	26	27	18	6
75	23	36	53	51	35	30	59	7
74	24	50	40	45	38	32	58	8
9	25	33	39	41	43	49	57	73
10	60	34	44	37	42	48	22	72
14	63	52	29	31	47	46	19	68
15	64	17	20	21	56	55	54	67
80	1	3	4	5	69	70	71	66

Ninth-order Frénicle bordered square

The 9×9 square meets the standard requirement that all rows, columns, and main diagonals add up to the magic sum 369. The internal 7×7 square is a magic square with a magic sum of 287. The fifth-order magic square has a magic sum of 205. The third-order square has magic sum 123.

Borders are numbered starting from the outside moving in to the center. For example, the first border is the outside border, the second border is the border next to the outside border, and so forth. Various references in the literature give recipes for creating bordered magic squares.[9]

Composite (or Compound) Squares

A *composite* or *compound magic square* is a magic square composed of a number of smaller magic squares that don't necessarily start at 1.

If you know how to make a square of the Mth and Nth orders, you can easily make one of the MNth order as shown in this example, where M and N are each 3. The composite magic square has the magic sum $9(81 + 1) / 2 = 369$.

Take a close look. This 9×9 square is composed of nine 3×3 magic squares. You can see that each 3×3 subsquare is successively constructed in the same order as the square in bold at the top, which contains the smallest integers. In other words, each successive subsquare is placed in the larger square in the same order as the numbers of the internal square. For example, the second 3×3 square to be filled is the shaded one at lower right, just as the 2 is at the lower right of the first subsquare. This reminds me of a fractal, a geometrical pattern that has similar structures within larger structures.

71	64	69	8	1	6	53	46	51
66	68	70	3	5	7	48	50	52
67	72	65	4	9	2	49	54	47
26	19	24	44	37	42	62	55	60
21	23	25	39	41	43	57	59	61
22	27	20	40	45	38	58	63	56
35	28	33	80	73	78	17	10	15
30	32	34	75	77	79	12	14	16
31	36	29	76	81	74	13	18	11

By coloring all the odd cells black and all even cells white, or by coloring values evenly divisible by 3, we can visualize additional patterns. I like to think of the coloring process as applying a stain to a wood grain to visually bring out its hidden, subtle structure.

	64		8		6		46	
66	68	70				48	50	52
	72		4		2		54	
26		24	44		42	62		60
22		20	40		38	58		56
	28		80		78		10	
30	32	34				12	14	16
	36		76		74		18	

Coloring odd cells black

71	64	69	8	1	6	53	46	51
66	68	70	3	5	7	48	50	52
67	72	65	4	9	2	49	54	47
26	19	24	44	37	42	62	55	60
21	23	25	39	41	43	57	59	61
22	27	20	40	45	38	58	63	56
35	28	33	80	73	78	17	10	15
30	32	34	75	77	79	12	14	16
31	36	29	76	81	74	13	18	11

Coloring divisible-by-3 cells gray

You can visualize the creation of this magic square in other ways. For example, start with the first 3×3 square and create eight other squares by adding 9 to every cell, thus obtaining magic third-order squares with magic sums 42, 69, 123, 159, 177, 204, and 231. Then the squares are jammed together in the same order as the numbers in the initial magic square.

In the following example, we have a magic square of order 8 with magic sum 260. I really get a kick out of this awesome beauty. If we add all the numbers in every 2×2 subsquare, we get 130. For example, the upper left subsquare is highlighted, and $7 + 53 + 12 + 58 = 130$. (Although this is not really a composite square because the small subsquares are not magic, I've seen a few books that refer loosely to these kinds of square as composite.) Moreover, if we square every number in the 8×8 square, we get another magic square with a magic sum of 11,180.

7	53	41	27	2	52	48	30
12	58	38	24	13	63	35	17
51	1	29	47	54	8	28	42
64	14	18	36	57	11	23	37
25	43	55	5	32	46	50	4
22	40	60	10	19	33	61	15
45	31	3	49	44	26	6	56
34	20	16	62	39	21	9	59

"Composite" square with sixteen 2×2 subsquares

Overlapping Magic Squares

Like the composite or compound magic squares in the previous section, overlapping magic squares are composed of a number of smaller magic squares. However, with overlapping magic squares, the subsquares or "panels" are of various sizes and intersect with one another.

Here is a beautiful ninth-order overlapping magic square. The 9×9 square has magic constant 369. The two 4×4 subsquares have magic constant 169. The overlapping 5×5 subsquares have magic sum 205.

75	53	11	25	14	65	48	42	36
10	26	74	54	49	43	32	15	66
71	57	7	29	33	16	67	50	39
8	28	72	56	68	46	40	34	17
52	69	13	30	41	35	18	64	47
12	27	38	51	77	80	20	3	61
37	59	76	9	24	4	60	81	19
73	6	23	45	58	79	21	2	62
31	44	55	70	5	1	63	78	22

Ninth-order square with overlays

I enjoy reading D. F. Savage's 1917 description[10] of these odd-order squares containing a pair of subsquares overlapping by a single cell:

> The major squares are like those once famous Siamese twins, Eng and Chang, united by a vinculum, an organic part of each, through which vital currents must flow; the central cell containing the middle term 41, must be their bond of union, while it separates the other pair.

Other, more exotic overlapping squares are exhibited in chapter 3.

Misfits or Imperfect Magic Squares

Although a magic square traditionally includes all the numbers from 1 up to N^2, this condition is sometimes relaxed to permit all kinds of fascinating misfits. To prepare your mind for the more intricate structures in the Gallery sections, consider some simple misfits such as the following square that uses consecutive *odd* numbers from 1 to 17.

15	1	11
5	9	13
7	17	3

Third-order misfit

As alluded to in the previous chapter, there are many ways we can create a third-order square using three consecutive runs of integers, such as (1, 2, 3), (7, 8, 9), and (13, 14, 15). Consider this example:

14	1	9
3	8	13
7	15	2

Third-order misfit

The magic sum for this square is 24. Sometimes when you construct these kinds of squares, you'll discover that the horizontal sums are all the same, and the vertical sums are all the same, but the horizontal sums are not necessarily equal to the vertical sums. You can experiment with large numbers and longer runs of numbers, such as the following square that is constructed from (1, 2, 3, 4), (6, 7, 8, 9), (11, 12, 13, 14), and (16, 17, 18, 19):

1	17	8	14
18	4	11	7
12	6	19	3
9	13	2	16

Fourth-order misfit

This misfit has a magic sum of 40. You can also play with magic squares composed of consecutive numbers starting with numbers other than 1, such as this magic square with numbers from 114 to 122 and magic sum of 354.

121	114	119
116	118	120
117	122	115

Third-order misfit

An imperfect magic square can always be constructed from an existing magic square by adding the same number to each of the numbers forming the magic square, or by multiplying or dividing each of the original numbers by a constant. For example, the following square on the right was created from the square on the left by dividing each number by 2.

1	15	14	4
12	6	7	9
8	10	11	5
13	3	2	16

$\mathcal{S} = 34$

½	7½	7	2
6	3	3½	4½
4	5	5½	2½
6½	1½	1	8

$\mathcal{S} = 17$

It is possible to compute the magic constant for these kinds of squares with simple formulas. When some number β is added to each number in the original square, then the magic sum for the new square is

$$\mathcal{S} = \mathcal{S}_0 + \beta N$$

where $\mathcal{S}$ is the new magic constant, $\mathcal{S}_0$ is the constant of the original square, N is the order of the square, and β is the number added to each number of the original square. When β is subtracted from each number of the original magic square, we get

$$\mathcal{S} = \mathcal{S}_0 - \beta N$$

When each number of the original square is multiplied by β, we get

$$\mathcal{S} = \beta \mathcal{S}_0$$

When each number of the original square is divided by β, we get

$$\mathcal{S} = C_0 / \beta$$

A new magic square Ω can be created from an original magic square ξ by adding the entries in $(N-1) \times (N-1)$ subsquares in the original.[11] As an example, consider the following third-order square ξ:

8	1	6
3	5	7
4	9	2

Magic square ξ

23	16	21
18	20	22
19	24	17

Magic square Ω

In this example, a new magic square Ω can be created from ξ by adding entries in 2×2 blocks. The (i, j) entry in Ω is obtained by deleting the elements in row i and column j of magic square ξ and adding the entries in the remaining block. For example, to get 23 in Ω we strike out row 1 and column 1 (because the upper left cell is in row 1 and column 1) and add the remaining cells in ξ. The "striking out" is represented in gray. Here is another example for an order-4 magic square.

16	2	3	13
5	11	10	8
9	7	6	12
4	14	15	1

ξ

84	70	71	81
73	79	78	76
77	75	74	80
72	82	83	69

Ω

In 1995, Emanuel Emanouilidis finally proved that Ω is always a magic square.[12] Notice that for third-order squares, the magic sum for Ω is four times the sum for ξ. For a fourth-order square, the magic sum for Ω is nine times the sum for ξ.

Subtracting, Multiplying, and Dividing Magic Squares

It is possible to consider squares that use subtracting, multiplying, or dividing in various interesting ways. For example, in the following subtracting square, you get a magic constant by subtracting the first number in a row from the second and the result from the third. To avoid negative numbers, you can deduct the middle number from the sum of the two end numbers, thereby yielding 5 for this square's constant. You can do this with the columns and diagonals as well.

2	1	4
3	5	7
6	9	8

Subtracting

12	1	18
9	6	4
2	36	3

Multiplying

3	1	2
9	6	4
18	36	12

Dividing

In multiplying magic squares, the products of all numbers in every horizontal row, vertical column, and main diagonal are the same. In this example of a multiplying magic square, you obtain the constant 216 by multiplying the three numbers in any row, column, or diagonal. In the dividing square shown here, you can obtain the constant 6 by dividing the second number in a row, column, or diagonal by the first in either direction and the third number by the quotient. To simplify, you can instead divide the product of the two extreme numbers by the middle number. Notice that the dividing square can be created from the multiplying square simply by reversing the order of the main diagonals of the multiplying square.

Knight's Move Magic Squares

To create a knight's move magic square, a chess knight has to be jumped once to every square on the (8 × 8) chessboard in a complete tour, with the squares visited numbered in order so that the magic sum is 260. One solution is shown here. Start at the 1 at the bottom, jump to the 2, and so forth.

46	55	44	19	58	9	22	7
43	18	47	56	21	6	59	10
54	45	20	41	12	57	8	23
17	42	53	48	5	24	11	60
52	3	32	13	40	61	34	25
31	16	49	4	33	28	37	62
2	51	14	29	64	39	26	35
15	30	1	50	27	36	63	38

♞ Jaenisch's knight's square ♞

Notice that the knight can jump finally from 64 back to 1, a beautiful feature of this square. If the first and last squares traversed are connected by a move, the tour is said to be *closed* (or *reentrant*); otherwise it is *open*.

This magic square was created in 1862 by C. F. Jaenisch in *Applications de l'Analyse Mathématique au Jeu des Echecs.* Sadly, it is not a perfect magic square because one diagonal sums to 264 and the other to 256. Therefore, the square is sometimes referred to as *semimagic.* I can imagine poor C. F. weeping—so close, yet so far!

For centuries, the holy grail of magic squares was to find a perfect knight's move magic square, but sadly all have had minor flaws. (It is possible to produce a knight's move magic square for larger boards, such as shown in Gallery 1 for a 16 × 16 board.) Aside from the Euler knight's square, all the examples in this section were discovered in the second half of the nineteenth century.

The first person to attempt to draw a knight's magic square was Leonhard Euler (1707–1783), a Swiss mathematician, and the most prolific mathematician in history.[13] Even when he was completely blind, Euler made great contributions to modern analytic geometry, trigonometry, calculus, and number theory.

1	48	31	50	33	16	63	18
30	51	46	3	62	19	14	35
47	2	49	32	15	34	17	64
52	29	4	45	20	61	36	13
5	44	25	56	9	40	21	60
28	53	8	41	24	57	12	37
43	6	55	26	39	10	59	22
54	27	42	7	58	23	38	11

♞ Euler's knight's square ♞

To traverse the square, start at the 1 in the upper left and leave the square at the 64 at right. Figure 2.5 shows the actual path. Notice the nice symmetries in the path. The right side of the figure is a mirror image of the left.

Like Ben Franklin's 8 × 8 magic square, which you'll encounter in the next chapter, the numbers in all 2 × 2 subsquares add up to

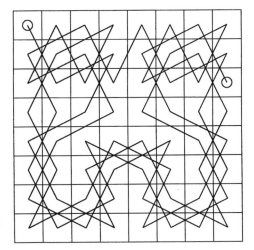

Figure 2.5
The path of a knight as it tours Euler's square.

130. Euler's knight's square adds up to 260 in the rows and columns. The four 4 × 4 subsquares are also magic in their rows and columns, which sum to 130. (The Euler knight's square is actually four 4th-order magic squares put together.) Alas, like the Jaenisch square, the main diagonals do not add to 260.

Euler was not merely a magic square expert. He published over 8000 books and papers, almost all in Latin, on every aspect of pure and applied mathematics, physics, and astronomy. He studied infinite series and differential equations. He also introduced many new functions (e.g., the gamma function and elliptic integrals) and created the calculus of variations. His notations such as e and π are still used today. In mechanics, he studied the motion of rigid bodies in three dimensions, the construction and control of ships, and celestial mechanics. Leonhard Euler was so prolific that his papers were still being published for the first time two centuries after his death. His collected works have been printed bit by bit since 1910 and will eventually occupy more than seventy-five large books.

The magic square of Feisthamel fulfills the requirement of adding up to 260 in rows, columns, and diagonals. But still it is not a true knight's move magic square. Can you tell why? After starting at 1, it turns out that the knight must come to a halt at square 32 and then slide like a rook horizontally to 33 before continuing its standard jumping pattern. Notice that the 1 and 64 are in the same row.

5	14	53	62	3	12	51	60
54	63	4	13	52	61	2	11
15	6	55	24	41	10	59	50
64	25	16	7	58	49	40	1
17	56	33	42	23	32	9	48
34	43	26	57	8	39	22	31
27	18	45	36	29	20	47	38
44	35	28	19	46	37	30	21

♞ Feisthamel's knight's square ♞

Also notice that cells on opposite sides of a vertical mirror plane always add up to 65. For example, in the top row we have 62 + 3, 53 + 12, 14 + 51, and 5 + 60.

The Wenzelides square is semimagic; the rows and columns sum to 260, but the diagonals add up to 192 and 328. The path of the knight is continuous from 1 to 64, with a nice finishing move where the knight jumps from 64 back to 1. The square exhibits a form of skew or radial symmetry in which numbers on each side of a diagonal line drawn through the midpoint have a difference of 32. For example, examining the center squares, we find 52 − 20 = 32 and 33 − 1 = 22.

It seems one can never find a Nasik 8 × 8 knight's move magic square. We know that the magic constant for an order-8

50	11	24	63	14	37	26	35
23	62	51	12	25	34	15	38
10	49	64	21	40	13	36	27
61	22	9	52	33	28	39	16
48	7	60	1	20	41	54	29
59	4	45	8	53	32	17	42
6	47	2	57	44	19	30	55
3	58	5	46	31	56	43	18

♞ Wenzelides knight's square ♞

square is 260. If a knight's movement starts from a black cell on a chessboard, it must land on a white cell. In this case, all black cells would be occupied by odd numbers and all white cells by even numbers. This means that each major diagonal and pandiagonal is composed of either all even or all odd numbers. We know that the sum of odd numbers $1 + 3 + 5 + \cdots + 63 = 1024$, which is less than 260×4. The sum of the even numbers $2 + 4 + 6 + \cdots + 64 = 1056$ is greater than 260×4. Therefore the Nasik square condition (that the sums be constant for all pandiagonals and major diagonals) cannot be satisfied.

Chess player and artist Ronald R. Brown from Pennsylvania composes music using the chess knight's tour.[14] If you don't worry about hopping along magic squares, the number of distinct solutions to the knight's tour problem on a chessboard are immense—estimates range from 31 million to 168! / 105!63! (The exclamation point is the factorial sign: $n! = 1 \times 2 \times 3 \times \cdots \times n.$) To create chess music, Brown first writes a solution to the knight's problem, such as in the following Wenzelides square:

50	11	24	63	14	37	26	35	Up 4
23	62	51	12	25	34	15	38	Up 3
10	49	64	21	40	13	36	27	Up 2
61	22	9	52	33	28	39	16	Up 1
48	7	60	1	20	41	54	29	Middle C
59	4	45	8	53	32	17	42	Down 1
6	47	2	57	44	19	30	55	Down 2
3	58	5	46	31	56	43	18	Down 3

♪ Making music with the knight's square ♪

Again, to understand this table of numbers, the knight starts at the position marked 1 and then proceeds to the position marked 2 and so on, traversing all the squares on the chessboard. This can be mapped to pleasing music by considering each knight's position as a note, the pitch of which is determined by the knight's row. Starting at middle C, the next note is two white notes lower, the third note three white notes lower (from middle C), and so on. By tracing various paths that the knight follows as it meanders around the board, Ronald Brown also produces interesting abstract art. A recent newspaper article quotes Brown describing his chess art: "Some people don't approve of this because they feel art must be spontaneous. My answer is that this is spontaneous because I don't know what it's going to look like until I've done it."[15]

Magic Stars

A *magic star* is a variation of a magic square. Numbers are arranged in a star formation such that the sum of the numbers in each of the straight lines formed by the star's corners and intersections is constant. The number of integers required to form a magic star is equal to two times the number of points in the star. Consecutive

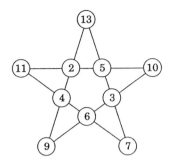

Figure 2.6
Magic pentagram.

integers are not always possible in forming a magic star. The five-pointed star is the smallest magic star that can be made. Ten integers are required to form this star, and there is no way to form a five-pointed star using consecutive integers. In the magic five-pointed star, or pentagram, in Figure 2.6, the sum of the numbers in every line is 28.

There are 479,001,600 ways of arranging consecutive numbers from 1 to 12 at the nodes of a six-pointed star, or hexagram, shown in Figure 2.7. Of these millions of possibilities, there are 80 magic hexagrams[16] that can each be shown in twelve ways. The sum of the numbers in every line is 26.

Magic Circles and Spheres

A *magic circle* is an arrangement of numbers in circles such that the sum of the numbers in each diameter is equal to the sum of the numbers in any two radii. Other kinds of magic circles are exhibited in chapter 4.

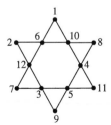

Figure 2.7
Magic hexagram.

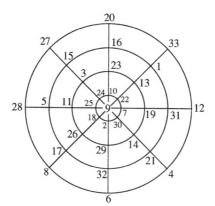

Figure 2.8
Yang Hui's magic circles.

The magic circles in Figure 2.8 were created in 1275 by Yang Hui, a minor Chinese official who wrote two books, dated 1261 and 1275, that used modern decimal fractions and gave the first account of Pascal's triangle. The 1275 work is called *Cheng Chu Tong Bian Ben Mo* (*Alpha and Omega of Variations on Multiplication and Division*). One of the more remarkable aspects of this work is the preface on mathematics education, "Xi Suan Gang Mu" ("A Syllabus of Mathematics"). Man Keung Siu[17] later writes that the syllabus

> . . . is an important and unusual extant document in mathematics education in ancient China. Not only does it specify the content and the time-table of a comprehensive study program in mathematics, it also explains the rationale behind the design of such a curriculum. It emphasizes a systematic and coherent program that is based on real understanding rather than on rote learning. This program is a marked improvement on the traditional way of learning mathematics by which a student is assigned certain classical texts, to be studied one followed by the other, each for a period of one to two years!

In Yang Hui's magic circles, the sum of any circle is 138, and the sum of any diameter is also 138 if we do not count the central 9.

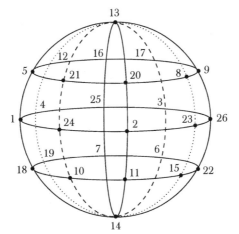

Figure 2.9
Magic sphere.

The magic sphere in Figure 2.9 has five great circles (equator and four double meridians) and two small (latitude) circles on each side of the equator, creating 26 nodes or crossing points. We can fill these nodes with the numbers 1 to 26 so that eight numbers on every circle add up to 108. Diametrically opposite numbers have the sum of 27–for example, the 2 toward the viewer and the 25 directly opposite. For clarity, larger numbers and grid dots in the figure are points toward the viewer.

Magic Cubes

A *magic cube* is a cube composed of N^3 numbers, where N is the order of one of the cube's square cross sections. The numbers used in forming a *perfect magic cube* are usually 1 through N^3. They are arranged so that each row, each column, each main diagonal of the square cross sections, and each of the four great (space) diagonals (sometimes called "triagonals") containing N integers will add to the same sum. The constant sum to which each row, column, diagonal, and great diagonal adds occurs in $3N^2 + 4$ ways. The magic constant is:

$$\mathcal{S} = \tfrac{1}{2}(N^4 + N)$$

A *simple magic cube* or *semiperfect magic cube* meets just the minimum requirements that the sums of the numbers in each row, column, and great diagonal are equal. However, like magic squares, various features may be added to such a cube to make it more magical, thus elevating it to a zenlike mandala worthy of long contemplation. A few of these intricate three-dimensional arrangements are exhibited in chapter 3.

Magic cubes can be constructed of odd or doubly even orders by extending the methods used in chapter 1. Like many magic square examples, there are no standard rules for constructing magic cubes of singly even order.

There is a unique perfect cube of order 1. There is no perfect cube of order 2, 3, or 4.

Martin Gardner has presented Lewis Myer's proof that there is no $3 \times 3 \times 3$ perfect cube.[18] Consider any 3×3 cross section. Let A, B, C be the numbers of the first row, D, E, F the numbers of the third, and X the central number. The magic constant for this cube is $\frac{1}{2}(3^4 \times 3) = 42$.

A	B	C
	X	
D	E	F

Cube's cross section

We know that the right diagonal sum is 42 ($C + X + D = 42$), the left diagonal sum is 42 ($A + X + F = 42$), and middle column sum is 42 ($B + X + E = 42$). Adding these three equations together, we get

$$3X + A + B + C + D + E + F = 3 \times 42 .$$

We also know that $A + B + C + D + E + F = 2 \times 42$. Subtract this formula from the previous to get $3X = 42$, or $X = 14$. However, the magic cube must be formed of consecutive, nonrepeating numbers. Because $X = 14$ cannot be the center of every cross section, the cube is impossible.

Magic cube junkies, understandably disturbed that no perfect 3×3 cube exists, relaxed the definition to create semiperfect magic cubes that do exist in all orders higher than 2. (As discussed, these cubes are such that only the rows, columns, and four space diagonals are magic.) W. S. Andrews devotes many pages to their description in his 1917 book *Magic Squares and Cubes*.[19] The order-3 Andrews semiperfect cube must be associate, with 14 in its center. There are four such cubes, not counting reflections and rotations.

No perfect cube of order 4 exists, as demonstrated by Richard Schroeppel, a mathematician and computer programmer.[20] The first step is to show that on any 4×4 section (orthogonal or diagonal), the four corners must add up to the magic constant. Let ⊠ be the magic constant, and label the sixteen cells of the cross section with letters as shown here.

Cross section of a fourth-order cube

The arrows indicate the paths for six lines (row 1, row 4, column 1, column 4, and two diagonals) that intersect all sixteen cells.

Row 1: $A + B + C + D = \boxtimes$

Row 4: $M + N + O + P = \boxtimes$

Column 1: $A + E + I + M = \boxtimes$

Column 4: $D + H + L + P = \boxtimes$

Right diagonal: $M + J + G + D = \boxtimes$

Left diagonal: $A + F + K + P = \boxtimes$

Because each corner cell is common to three lines, we get

$3A + 3D + 3M + 3P +$ each other cell taken once $= 6 \times \boxtimes$

Let's subtract out the values of the four rows, to get

$$2A + 2D + 2M + 2P = 2\boxtimes$$

or

$$A + D + M + P = \boxtimes$$

This proves that the four corners of any cross section add up to the magic constant.

Now we consider the cube's eight corner cells and prove that any two corners connected by an edge must have a sum of $\boxtimes / 2$. The following are four cross sections of the cube. Let us choose the two corners A and B. Let C-D and E-F be the corners of two edges parallel to A-B.

First cross
section

Second
cross
section

Third cross
section

Fourth cross
section

The cells *A, B, C, D*; *E, F, B, A*; and *E, F, D, C* are each the corners of a different 4 × 4 cross section, so that their total is 3⊠. Gathering like terms, we get $2A + 2B + 2C + 2D + 2E + 2F = 3⊠$. Divide each side by 2 to get $A + B + C + D + E + F = 3⊠/2$. Because we just showed that the four corners of any cross section sum to ⊠, from this last equation, subtract $C + D + E + F = ⊠$ to get

$$A + B = ⊠/2$$

We have just proved that any two corners connected by an edge must have the sum of half the magic constant.

Now consider corner *B*. It is connected through a straight line of cells to corners *A, D*, and *F*. Because $A + B = F + B = D + B$, we can take *B* from each equality to prove that $A = F = D$. But this is impossible for a perfect magic cube with consecutive numbers in the cells, so we have proved that no perfect fourth-order cube exists.

Did you feel good after that little mental workout?

Through every number in a perfect magic cube, one can go in thirteen possible directions if one considers the rows, columns, and various diagonals. We've just shown that in a third- or fourth-order cube, these thirteen sums can't be the same. You'll have to go to an order-8 cube before this becomes possible. Figure 2.10 shows a third-order magic cube with seven of the thirteen different directions drawn for the center element 14. Can you draw the remaining six directions? Each of these thirteen lines through the center sums to 42.

Although there are no perfect magic cubes of order less than 8, there are millions of perfect order-8 magic cubes,[21] an example of which is exhibited in Gallery 1. There are also perfect magic cubes of order 64 and 512 and other powers of 8. In 1962, H. Langman constructed what he considered to be the first perfect magic cube of order 7; however, such cubes appear to have repetitions along one of their diagonals or triagonals.[22] Some have conjectured that an order-10 perfect magic cube might exist, but I do not know if one has been found.

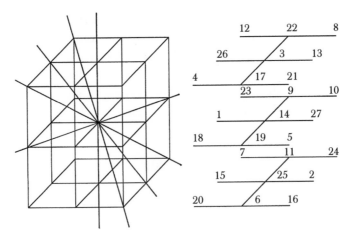

Figure 2.10
Thirteen different lines pass through the center (14) of this third-order magic cube, and all the lines sum to 42. Seven of these directions are shown on the left. Can you draw the remaining six? At right is a schematic of the cube showing the positions of the twenty-seven numbers that fill the cube's cells. [After John Hendricks, Magic Squares to Tesseracts by Computer, *self-published, 1998, 52.]*

The cube in Figure 2.11 was published in the late 1800s by the Chinese Pao Chhi-shou in his *Pi Nai Shan Fang Chi* ("Pi Nai Mountain Huts Records"). This is a different kind of magic cube because, as we discussed, magic cubes of the Nth order usually have N^3 numbers, but Chhi-shou developed this ingenious cube with thirty-two numbers so that all edges sum to 41 *between* the vertices. Today

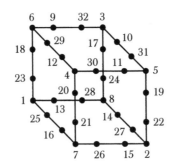

Figure 2.11
Pao Chhi-shou's magic cube.

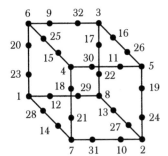

Figure 2.12
Modified Pao cube.

we can do a bit better than Chhi-shou and keep the position of his first eight numbers the same but rearrange the remainder to create a figure in which the sum along every edge, including vertices, is 50 (Fig. 2.12).

If replacing each number in a magic cube by its square produces another magic cube, the magic cube is said to be a *bimagic cube*. In June 2000, John Hendricks created the first bimagic cube. Hendricks's bimagic cube was of order 25 and contained the consecutive numbers from 1 to 15,625. The magic sum is 195,325. When you square all the numbers in the cube, the magic sum becomes 2,034,700,525. To check his calculations, Hendricks used a small computer to determine the sums along rows, columns, pillars, diagonals, and triagonals. Any reader interested in the details of Hendricks's methods may write to me.

While on the topic of magic cubes and their diagonals, I can't help mentioning one of the holy grails of numerical cubes: the solution to the "integer brick problem." Here one must find the dimensions of a three-dimensional brick such that the distance between any two vertices is an integer (Fig. 2.13). In other words, you must find integer values for a, b, and c (which represent the lengths of the brick's edges) that produce integer values for the various diagonals of each side: d, e, and f. In addition, the three-dimensional space diagonal (triagonal) g spanning the brick must also be an integer. This means that the following equations must

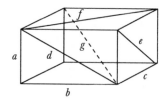

Figure 2.13
The fiendishly difficult
"integer brick" problem.

have an integer solution:

$$a^2 + b^2 = d^2$$

$$a^2 + c^2 = e^2$$

$$b^2 + c^2 = f^2$$

$$a^2 + b^2 + c^2 = g^2$$

No solution has been found. However, mathematicians haven't been able to prove that no solution exists. Many solutions have been found with only one noninteger side.

Pythagorean Magic Squares

In order to understand Pythagorean magic squares, I relish the excuse to digress and review Pythagoras's famous theorem that in a right-angled triangle (Fig. 2.14) the sum of the squares of the shorter sides, a and b, are equal to the square of the hypotenuse c; that is, $c^2 = a^2 + b^2$. With this knowledge, we can always determine the third side of a right triangle when the other two sides are

Figure 2.14
A Pythagorean triangle.

known. More proofs have been published of Pythagoras's theorem than of any other proposition in mathematics! There have been several hundred proofs.

Pythagorean triangles are ones where *a-b-c* are integers, such as 3-4-5 and 5-12-13. Pythagoras's favorite, 3-4-5, has a number of properties not shared by other Pythagorean triangles, apart from its multiples such as 6-8-10. It is the only Pythagorean triangle whose three sides are consecutive numbers. It is the only triangle of *any* shape with integer sides the sum of which (12) is equal to twice its area (6).

Here's something that may make you think twice about 666, the number of the Beast in the Book of Revelation in the Bible. There exists only one Pythagorean triangle, except for the 3-4-5 triangle, whose area is expressed by a single digit. It's the triangle 693-1924-2045, having area

$$\boxed{666,666}$$

Pythagorean magic squares are sets of three magic squares positioned on each side of a right triangle such that the sum of the squares of the numbers in the magic square on the hypotenuse equals the sum of the squares of the numbers in the magic squares on the other sides. Figure 2.15 shows an example of such a set.

To create this set of squares, start by drawing a right triangle. Draw two squares that sit on the legs *a* and *b* of the triangle and

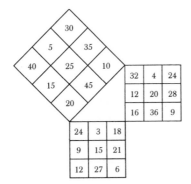

Figure 2.15
A Pythagorean magic
square.

another square that sits on the hypotenuse *c*. The squares can be of any order you like, as long as they are the same order. In this example, we use third-order squares arranged along a triangle where $a = 4$, $b = 3$, and $c = 5$. In order to fill the squares, we have to use a specific series of numbers. The numbers to use in magic square *c* are *c*, $2c$, $3c$, . . . , cN^2, where *c* is the hypotenuse length and *N* is the order of the square. Similarly, square *b* is filled with *a*, $2a$, $3a$, . . . , aN^2. Square *b* is filled with *b*, $2b$, $3b$, . . . , bN^2.

In our example, square *c* is filled with the numbers 5, 10, 15, 20, 25, 30, 35, 40, 45. Square *a* is filled with the numbers 4, 8, 12, 16, 20, 24, 28, 32, 36. Square *b* is filled with 3, 6, 9, 12, 15, 18, 21, 24, 27. For each square, arrange the numbers by any method appropriate for constructing magic squares of odd order. For this example, I used de la Loubère's method.

You now have a beautiful Pythagorean set of three magic squares with some interesting properties:

- Select any three corresponding cells of squares *a*, *b*, and *c*, for example, 24, 32, and 40 in the upper left-hand corners. Any such cells will follow the relation $c^2 = a^2 + b^2$.
- Select any two cells in square *c*. Add the numbers and square the sum. Do the same for the corresponding squares *a* and *b*. You will find that the sums also follow the Pythagorean relation.

  Square *c*: $40 + 5 = 45$
  Square *a*: $32 + 4 = 36$
  Square *b*: $24 + 3 = 27$
  Square the sums to get: $45^2 = 36^2 + 37^2$

- Select any row, column, or diagonal in square *c*. Square each number and then add the results together. Do the same for the corresponding numbers in squares *a* and *b*. The sum for *c* will equal the sum for *a* plus the sum for *b*.
- Select any row, column, or diagonal in square *c*. Square the sum of the numbers. Do the same for the corresponding numbers in squares *a* and *b*. The squared sum for *c* will equal the squared sum for *a* plus the squared sum for *b*.

- Add all the numbers in square *c*, and square the sum. Do the same thing for squares *a* and *b*. The squared sum for *c* is equal to the squared sum for *b* plus the squared sum for *a*.
- Square all the numbers in square *c*, and add the results. Do the same thing for squares *a* and *b*. The sum for *c* equals the sum for *b* plus the sum for *a*.

All of these results are true for higher-order Pythagorean sets constructed in a similar manner.

Classification by Center Number

As you have seen throughout this chapter, the task of classifying magic squares in meaningful ways is very challenging. Certain categorizations may seem as useful as dividing the human race into people who eat sushi and those who do not. Nevertheless, certain divisions yield unexpected results. For example, Martin Gardner[23] enjoys looking at the total number of order-5 squares whose centers contain the numbers 1 through 13:

Center Number	Number of Magic Squares	Center Number	Number of Magic Squares
1	1,091,448	8	3,112,161
2	1,366,179	9	3,472,540
3	1,914,984	10	3,344,034
4	1,958,837	11	3,933,818
5	2,431,806	12	3,784,618
6	2,600,879	13	4,769,936
7	3,016,881		

One way to classify order-5 magic squares

Notice that the totals increase from 1 to 9, but then fluctuate from 9 to 13. Mathematicians were surprised that there are more squares with a center of 11 than squares with a center of 12, and more squares with a center of 9 than squares with a center of 10. These trends are echoed in counts of squares with centers 14 through 25 because every square with a center that is not 13 has a complement. There are as many squares with 1 in the center as there are with 25, and the same is true for all numbers except 13.

Antimagic Squares and Heterosquares

Heterosquares are $N \times N$ arrays of integers from 1 to N^2 such that the rows, columns, and main diagonals have *different* sums. These kinds of squares were first considered with enthusiasm in the early 1950s.[24] Royal V. Heath, an American magician and puzzle enthusiast, was first to prove that a 2×2 heterosquare formed with the numbers 1, 2, 3, and 4 is impossible. Consider that such a square would require six different sums for the two columns, two rows, and two diagonals. However, there are only five possible sums, the lowest being $3 = 1 + 2$ and the highest being $7 = 3 + 4$. Therefore, an order-2 heterosquare is impossible. Royal Heath also believed that an order-3 heterosquare was impossible, until Charles W. Trigg, the famous recreational mathematician, discovered the following square:

9	8	7
2	1	6
3	4	5

Heterosquare

Here are two methods for producing heterosquares. Start by writing the numbers from 1 to N^2 in order, as in the example here for the 4×4 square on the left.

1	2	3	4
5	6	7	8
9	10	11	12
13	14	15	16

Almost heterosquare

1	2	3	4
5	6	7	8
9	10	11	12
13	14	16	15

Heterosquare

In the square on the left, all rows and columns have different sums, but the two main diagonals have the same sum, 34. No problem. Just swap the integers 15 and 16 to produce the heterosquare on the right. Try this kind of approach for higher-order squares.

Another way to create an instant heterosquare is to start at the center with 1 and then spiral outward with consecutive integers. Here is an example for an order-9 heterosquare.

81	80	79	78	77	76	75	74	73
50	49	48	47	46	45	44	43	72
51	26	25	24	23	22	21	42	71
52	27	10	9	8	7	20	41	70
53	28	11	2	1	6	19	40	69
54	29	12	3	4	5	18	39	68
55	30	13	14	15	16	17	38	67
56	31	32	33	34	35	36	37	66
57	58	59	60	61	62	63	64	65

Spirogyra order-9 heterosquare

Trace the path round and round until you get dizzy. Make a poster showing a huge 51 × 51 version. Hang it on your wall. When friends ask, "So what?" just reply, "It's a spirogyra order-51 heterosquare," and watch their mouths drop in awe.

Does it seem a bit too easy to create a heterosquare? Don't fret. *Antimagic squares* are more challenging. An antimagic square is an $N \times N$ array of numbers from 1 to N^2 such that the rows, columns, and main diagonals produce different sums, and the *sums* form a consecutive series of integers. J. A. Lindon, an English puzzle aficionado, did much of the pioneering work on antimagic square construction, which you can read about in Joseph Madachy's *Madachy's Mathematical Recreations.*[25] Although there are many ways to create magic squares (as you saw in the previous chapter), there seem to be almost no simple systematic ways for creating antimagic squares. I would be interested in receiving any recipes from readers.

I wonder if there is a way to convert magic squares to antimagic squares by some simple recipe. What strange symmetries might we someday discover in antimagic squares that are comparable to the symmetries in magic squares?

Below is a square that is almost antimagic because the row sums are 16, 18, and 11; the column sums are 13, 15, and 17; and the diagonal sums are 12 and 15. The problem: 15 is repeated, and there is no sum of 14.

1	8	7
9	5	4
3	2	6

Almost antimagic square

It seems that antimagic squares of orders 1, 2, and 3 are impossible, but higher orders occur. For example, here is a fourth-order antimagic square.

6	8	9	7
3	12	5	11
10	1	14	13
16	15	4	2

Antimagic square

Talisman Squares

Talisman squares were invented by Sidney Kravitz, a mathematician from Dover, New Jersey. A talisman square is an $N \times N$ array of numbers from 1 to N^2 in which the difference between any one number and its neighbor is greater than some given constant. A *neighboring number* is defined as being in a cell horizontally, vertically, or diagonally adjacent to the current cell.

On the left is an example of a talisman square in which the difference between any number and its neighbor is greater than 1. The number 1, for example, has three neighbors (5, 9, and 11). The number 11 has eight neighbors (1, 5, 3, 13, 4, 5, 2, and 9). On the right is a 5 × 5 talisman square with differences greater than 4.

1	5	3	7
9	11	13	15
2	6	4	8
10	12	14	16

Fourth-order talisman

15	1	12	4	9
20	7	22	18	24
16	2	13	5	10
21	8	23	19	25
17	3	14	6	11

Fifth-order talisman

Talisman squares have been studied only since the late 1970s, and no rules for constructing them are known. There do not seem to be theories that help determine the maximum possible difference between numbers and their neighbors. (If you attempt a third-order talisman with integers from 1 to 9, notice that it is impossible to have a difference greater than 1 between all neighboring integers.)

Fractional and Negative Magic Squares

As discussed in the Introduction, just as with traditional magic squares, a compact formula exists to calculate the magic constant for magic squares that start at numbers other than 1 and that make use of an arithmetic series with a constant difference between successive integers. The magic constant for these kinds of squares depends on the order N, the starting integer A, and the difference D between successive terms:

$$ S = N\left(\frac{2A + D(N^2 - 1)}{2}\right) $$

The equation can be transposed so that it is possible to calculate the starting integer A given the magic constant, difference, and order:

$$ A = \frac{S - DK}{N} $$

where

$$ K = \frac{N}{2}(N^2 - 1) $$

For example, we can use this formula to find the initial number required for a third-order square with 1 as the difference to produce a magic constant of 1903. The formula yields $[1903 - (1 \times 12)] / 3$. This means the initial number is fractional: $630\frac{1}{3}$.

$637\frac{1}{3}$	$630\frac{1}{3}$	$635\frac{1}{3}$
$632\frac{1}{3}$	$634\frac{1}{3}$	$636\frac{1}{3}$
$633\frac{1}{3}$	$638\frac{1}{3}$	$631\frac{1}{3}$

Magic square with fractions

Although fractional numbers are not traditionally used in magic squares, they make for interesting explorations. You can also transpose the formula to build squares using any desired initial number.

For example, you can use the formula

$$D = \frac{S - AN}{K}$$

to compute what difference must be used in a third-order square starting with the number 1 and with 1903 as the magic constant. We find that the difference is $\frac{1903 - 1 \times 3}{12} = 158\frac{1}{3}$. Starting with 1, we can construct the following square:

$1109\frac{1}{3}$	1	$792\frac{2}{3}$
$317\frac{1}{3}$	$634\frac{1}{3}$	951
476	$1267\frac{2}{3}$	$159\frac{1}{3}$

Magic square with fractions

Similarly, you can use these equations in some odd instances to yield magic squares with negative numbers. For example, if you try to compute the increment that must be used in a 4 × 4 square with starting number 48 and magic constant 42, you will generate the following square:

48	−22	−17	33
−7	23	18	8
13	3	−2	28
−12	38	43	−27

Magic square with negatives

I used only small squares as examples, but these formulas apply to magic squares of any size.

Magic Tesseracts

Most of the magical figures of this book are confined to a plane, although magic cubes and spheres are occasionally discussed. However, there is no reason why we cannot consider the properties of higher-dimensional figures such as tesseracts, the four-dimensional analogues of cubes. Mathematician John Hendricks has created many such figures, an example of which is exhibited in Gallery 1.

Magic Tesseract Type I

One form of magic tesseract requires that numbers be placed only at the sixteen corners of the tesseract. In these embodiments, the magic tesseracts are related to diabolic squares in interesting ways.[26] Recall that diabolic squares remain diabolic under rotations, reflections, transfer of a row from top to bottom or vice versa, transfer of a column from one side to the other, and rearranging of cells according to the plan shown here:

1	8	13	12
14	11	2	7
4	5	16	9
15	10	3	6

Diabolic square

a	b	c	d
e	f	g	h
i	j	k	l
m	n	o	p

Initial pattern

a	d	h	e
b	c	g	f
n	o	k	j
m	p	l	i

Rearrangement

The relation of the diabolic square to the tesseract is demonstrated by transferring the sixteen cells of the diabolic square (at left) to the sixteen corners of a hypercube. This can be shown by the two-dimensional projection of the hypercube in Figure 2.16.[27] The antipodal pairs, which add up to 17, are the diagonally opposite corners of the hypercube. By rotating and reflecting the hypercube, it can be placed in exactly 384 different positions, each of which maps back to the plane as one of the 384 diabolic squares.

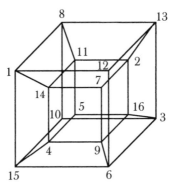

Figure 2.16
Magic tesseract, type 1.

If you're having trouble grasping the notion of hypercubes, you can get an idea about what they're like by starting in lower dimensions. For example, if you move a point from left to right, you trace out a one-dimensional line segment. Imagine drawing a line with chalk on a blackboard. If you take this line segment and move it up (perpendicularly) along the blackboard, you produce a two-dimensional square. If you move the square out of the blackboard, you produce a three-dimensional cube (Fig. 2.17).

I can hear you asking the question, "How can we move the square out of the blackboard?" The answer is that we can't do that, but we can graphically represent the perpendicular motion by moving the square—on the blackboard—in a direction *diagonal* to the first two motions. In fact, if we use the *other* diagonal direction to represent the fourth dimension, we can move the image of the cube in this fourth dimension to draw a picture of a four-

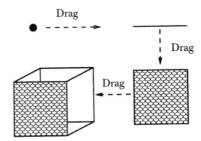

Figure 2.17
Lower-dimensional figures trace out higher-dimensional figures when the lower-dimensional figures are moved.

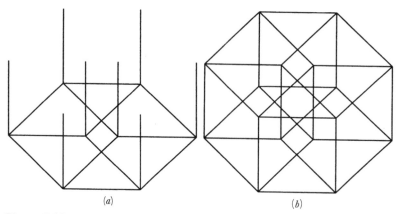

Figure 2.18
A hypercube (b) produced by moving (a) a cube along the fourth dimension.

dimensional hypercube, also known as a tesseract. Or we can rotate the cube and move it straight up in the drawing (Fig. 2.18). The tesseract is produced by the trail of a cube moving into the fourth dimension.

Our visual powers have a hard time moving the image of a cube, but we can assume that the cube is shifted a distance in a direction perpendicular to all three of its axes. We can even write down the number of corners, edges, faces, and solids for higher-dimensional objects.

Take a look at the hypercube drawing. Can you see the sixteen corners? The number of corners (or vertices) doubles each time we increase the dimension of the object. The hypercube has thirty-two edges. To get the volumes of each object, all you have to do is multiply the length of the sides. For example, the volume of a cube is l^3, where l is the length of a side. The hypervolume of a hypercube is l^4, the hyperhypervolume of a five-dimensional cube is l^5, and so on.

How can we understand that a hypercube has thirty-two edges? The hypercube can be created by displacing a cube and seeing the trail it leaves. Let's sum the edges. The initially placed cube and the finally placed cube each have twelve edges. The cube's eight

	Corners	Edges	Faces	Solids	Hyper-volumes
Point	1	0	0	0	0
Line segment	2	1	0	0	0
Square	4	4	1	0	0
Cube	8	12	6	1	0
Hypercube	16	32	24	8	1
Hyper-hypercube	32	80	80	40	10

corners each trace out an edge during the motion. This gives a total of thirty-two edges. The drawing is a nonperspective drawing, because the various faces don't get smaller the "farther" they are from your eye.

Imagine if I were to hand you a cube of sugar and a pin. Can you touch any point inside any of the square faces without the pin going through any other point on the face? The answer is yes. Now, think what that would mean for a hyperman in the fourth dimension touching the cubical "faces" of a tesseract. For one thing, a hyperman can touch any point inside any cubical face without the pin's passing through any point in the cube. Points are "inside" a cube only to you and me. To a hyperman, every point in each cubical face of a tesseract is directly exposed to his vision as he turns the tesseract in his hyperhands.

There's another way to draw a hypercube. Notice that if you look at a wire-frame model of a cube with its face directly in front of you, you will see a square within a square (Fig. 2.19). The smaller square is farther away from your eye and is drawn smaller because the drawing is a perspective drawing. If you looked at a hypercube in the same manner, you would see a cube within a

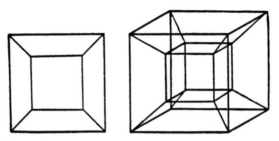

Figure 2.19
A wire-frame model of a cube viewed head-on and a tesseract.

cube. The closest part of the hypercube appears as a large cube, and the part farthest away appears as a smaller cube inside the larger one. This is called a *central projection* of the hypercube. More accurately, it is a plane projection of a three-dimensional model that is in turn a projection of a hypercube (Fig. 2.19). This is a shadow you might see if a hypercube is illuminated from a point "above" ordinary space in the fourth dimension.

Let me reiterate. A cube is bounded by square faces and a hypercube by cubical faces. A hypercube contains eight cubes on its hypersurface. But it's hard to see these eight cubes in the central-projection drawing (Fig. 2.19). Six of the eight cubes are distorted by projection, just as four of the cube's square faces are distorted when drawn on a plane. For a tesseract, the eight cubes are the large cube, the small interior cube, and the six hexahedrons (distorted cubes) surrounding the small interior cube. If you want your mind to be further shattered by 4-D concepts, see my book *Surfing Through Hyperspace: Understanding Higher Universes in Six Easy Lessons.*

Magic Tesseract Type II

In another form of magic tesseract, the figure contains the numbers 1 through N^4 arranged in such a way that the sum of the numbers in each of the N^3 rows, N^3 columns, N^3 pillars, N^3 "files" (a term used to imply a four-spatial direction), and in the eight major "quadragonals" (which pass through the center and join opposite

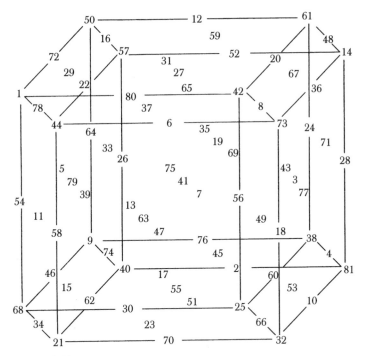

Figure 2.20
One of John Hendricks's third-order magic tesseracts.

corners) is a constant sum $\mathcal{S} = N(1 + N^4)/2$, where N is the order of the tesseract. Higher dimensions can also be imagined. For example, a five-dimensional magic hypercube of order 3 has the magic sum $\mathcal{S} = N(1 + N^5)/2 = 366$.

Figure 2.20 shows one of John Hendricks's third-order magic tesseracts.[28] Figure 2.21 shows an example of a row, column, pillar, and file sum. The twenty-seven rows, twenty-seven columns, twenty-seven pillars, and twenty-seven files must sum to 123.

Figure 2.22 shows how to locate the quadragonals. Here the opposite corners of the tesseract are shown by the same number. Quadragonals join opposite corners, and the numbers along the quadragonals must all sum to the magic sum. For example, in Figure 2.20 one quadragonal joins the opposite corners with 1 and 81 while passing through the central number 41. The sum is 123.

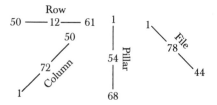

Figure 2.21
Example row, column, pillar, and file sum. The twenty-seven rows, twenty-seven columns, twenty-seven pillars, and twenty-seven files must sum to 123.

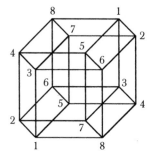

Figure 2.22
Guide for locating the quadragonals.

57 and 25 are also considered opposites. As there are sixteen corners, there are eight quadragonals.

There are fifty-eight different magic tesseracts of order 3. Each of them can be displayed in 384 ways due to rotations and/or reflections. This means that there are a total of 22,272 magic tesseracts of order 3.

In 1989 Joseph Arkin, David C. Arney, and Bruce J. Porter wrote an article about a $7 \times 7 \times 7 \times 7$ tesseract with magic sum 8407.[29] They named the tesseract the "Cameron cube" in honor of the recently retired Brigadier General David H. Cameron, who served in the Department of Mathematics at the United States Military Academy at West Point. At the time, the "Cameron cube" was thought to be the smallest known *perfect magic tesseract*. However, the term "perfect magic tesseract" implies that a magic sum is achieved not only in the rows, columns, pillars, files, and quadragonals but also in all the diagonals and triagonals (space diagonals of cubes).[30] By this criterion, the Cameron cube is not perfect, because, for example, one of the triagonals sums to $923 + 1917 + 559 + 1602 + 244 + 1238 + 2281 = 8764$, not 8407. A perfect magic tesseract requires all cubes to be perfect and all squares to be perfect (pandiagonal). John Hendricks has since proven that a perfect magic tesseract cannot be achieved with any orders below 16 and that a perfect magic tesseract of order 16 exists.[31] This perfect magic tesseract of order 16 contains the numbers $1, 2, 3, \ldots$, 65,536 and has the magic sum 534,296.

In April 1999 John and I computed the first perfect sixteenth-order magic tesseract. In 4-D space, we checked forty different directions along the various kinds of diagonals, rows, columns, etc. We used a systematic method to check the sums of sixteen numbers along each of these forty routes through each of the 65,536 points. There were approximately 2,621,440 computations. The total computation required about ten hours on an IBM IntelliStation running the Windows NT operating system. To assess the feasibility of running such calculations concurrently with normal office applications, such as e-mail and word processing, I ran the tasks at low priority with no perceived effect on the performance of the machine.

To summarize, we now know that the smallest perfect tesseract is of order 16, the smallest perfect cube is of order 8, and the smallest perfect (pandiagonal) magic square is of order 4.

Serrated Magic Squares

Serrated magic squares contain sawtooth edges, as shown at the top of the next page.[32] In serrated magic squares, the sums of the numbers in diagonals are the same. In this example, the diagonals are considered to consist of nine numbers such as $4 + 39 + 29 + 19 + 14 + 27 + 17 + 24 + 16 = 189$ or $4 + 20 + 41 + 12 + 28 + 27 + 33 + 18 + 6 = 189$. Additionally, all the "zigzag rows" and "zigzag columns" sum to 189. Zigzag and other magic patterns are shown in the following:

				14				
			19	16	30			
		29	24	40	8	1		
	39	17	5	36	11	25	22	
4	27	10	7	21	15	32	35	38
	20	33	31	26	37	9	3	
		41	18	2	34	13		
			12	6	23			
				28				

Serrated magic square

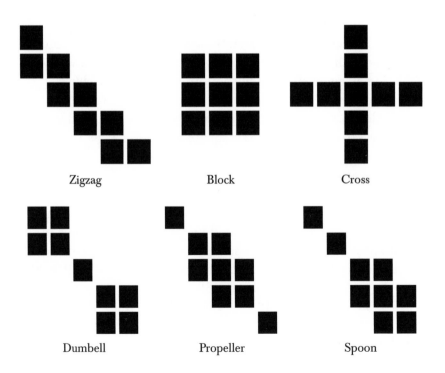

Zigzag   Block   Cross

Dumbell   Propeller   Spoon

In this particular serrated magic square, additional patterns sum to the magic sum. There are nine magic "crosses" and "blocks," six magic "dumbells" and "propellers," and twelve magic "spoons." Every other diagonal can be placed together to form a magic square. For example, (4, 39, 29, 19, 14), (20, 10, 5, 40, 30), (41, 31, 21, 11, 1), (12, 2, 37, 32, 22), and (28, 23, 13, 3, 38) form a 5 × 5 Nasik square. The remaining diagonals form a 4 × 4 Nasik square.

Other Wonderful and Famous Squares, Cubes, and Tesseracts

The classic puzzles in this section are not of the magic variety in which sums must be equal in various directions, but the puzzles seem apropos for this book because almost all mathematicians interested in magic squares seem to find delight in these other puzzles that are based on squares and cubes.

Fifteen Puzzle

Sam Loyd (1841–1911) has often been referred to as America's greatest puzzlist. The "14-15" puzzle, or just "fifteen puzzle," illustrated Loyd's interest in practical jokes. The puzzle is the equivalent of today's sliding-square puzzle with fifteen squares (tiles) and one vacant spot in a 4 × 4 frame or box. At startup, the squares sequentially contain the numbers 1 through 13, followed by 15, then 14. The idea is to "slide" the squares up, down, right, and left to arrive at the sequence 1 through 15. Loyd offered $1000 for the correct solution. Many people claimed to have solved the puzzle, but none

1	2	3	4
5	6	7	8
9	10	11	12
13	15	14	

Loyd's fifteen puzzle (starting position)

could duplicate their effort at collection time. Quite simply, they couldn't, and Loyd knew it because the puzzle is impossible to solve. In other words, the goal is to rearrange the squares from a given arbitrary starting arrangement (such as the previous one) by sliding squares one at a time into the configuration shown below. For some initial arrangements, this rearrangement is possible, but for others it is not.

1	2	3	4
5	6	7	8
9	10	11	12
13	14	15	

Loyd's fifteen puzzle (ending position)

This puzzle became an instant success, much like the Rubik's cube 100 years later. The only way to get from the standard starting position to the finishing position is to physically lift the 15 and 14 tiles out of the frame and then swap them, an illegal move. Can you imagine how many hours people must have spent attempting to solve this properly?

It's fun to think about what initial arrangements can lead to solutions. Here is the strategy.[33] If the sliding square containing the number i appears "before" (reading the squares in the box from left to right and top to bottom) n numbers that are less than i, then call it an inversion of order n, and denote it n_i. Then define

$$\gamma \equiv \sum_{i=1}^{15} n_i = \gamma \equiv \sum_{i=2}^{15} n_i$$

where the sum need run only from 2 to 15, rather than 1 to 15, because there are no numbers less than 1 (so n_1 must equal 0). If γ is even, the set of position is possible to create; otherwise, it is not. For example, in the following arrangement

2	1	3	4
5	6	7	8
9	10	11	12
13	14	15	

$n_2 = 1$ (2 precedes 1) and all other $n_i = 0$, so $\gamma = 1$, and the puzzle cannot be solved. I recall fondly, as a boy, giving a friend the above arrangement to solve, never telling the friend that it was impossible.

Famous Russian puzzlist Y. I. Perelman (1882–1942) quoted German mathematician W. Arens regarding Loyd's fifteen puzzle:[34]

About half a century ago, in the late 1870s, the Fifteen Puzzle bobbed up in the United States; it spread quickly and owing to the uncountable number of devoted players it had conquered, it became a plague. The same was observed on this side of the ocean, in Europe. Here you could even see the passengers in horse trams with the game in their hands. In offices and shops bosses were horrified by their employees being completely absorbed by the game during office and class hours. Owners of entertainment establishments were quick to latch onto the rage and organized large contests. The game had even made its way into solemn halls of the German Reichstag. "I can still visualize quite clearly the greyhaired people in the Reichstag intent on a small square box in their hands," recalls the geographer and mathematician Sigmund Gunter, who was a deputy during puzzle epidemic.

In Paris the puzzle flourished in the open air, in the boulevards, and proliferated speedily from the capital all over the provinces. A French author of the day wrote, "There was hardly one country cottage where this spider hadn't made its nest lying in wait for a victim to flounder in its web."

In 1880 the puzzle fever seems to have reached its climax. But soon the tyrant was overthrown and defeated by the weapon of Mathematics. The mathematical theory of the puzzle showed that of the many problems that might be offered, only half were solvable; the other half were impossible, however ingenious the technique applied to solve them.

It thus became clear why some problems would not yield under any conditions and why the organizers of the contests had dared offer such enormous rewards for solving problems. The inventor of the puzzle took the cake in this respect, suggesting to the editor of a New York newspaper that he publish an unsolvable problem in the Sunday edition with a reward of $1000 for its solution. The editor was a little reluctant, so the inventor expressed his willingness to pay his own money. The inventor was Sam Loyd. He came to be widely known as an author of amusing problems and a multitude of puzzles.

Interestingly enough, Loyd failed to patent his fifteen puzzle in the United States. According to patent law of the time, it seemed that he had to submit a "working model." Loyd posed the problem to a patent office official, who asked if the puzzle were solvable. When Loyd told him no, the official replied, "In which case there can't be a working model, and without a working model there can be no patent." Loyd appeared satisfied with the decision, but perhaps he should have been more tenacious, given the huge success of his invention.

Loyd himself remarked about his puzzle's popularity,

The $1000 reward offered for the first correct solution remained unretrieved although everybody was busy on it. Funny stories were told of shop-keepers who forget for this reason to open their shops, of respectful officials who stood throughout the night under a street lamp seeking a way to solve it. Nobody wanted to give up, as everyone was confident of imminent success. It was said that navigators allowed their ships to run aground, engine drivers took their trains past stations, and farmers neglected their ploughs.

Y. I. Perelman summarized the state of affairs thirty years after Loyd's death:

Thanks to the new light shed on the puzzle by Mathematics, the earlier morbid passion that was shown for the game is now unthinkable. Mathematics has produced an exhaustive explanation of the game, one that leaves no loophole. The outcome of the game is dependent not on chance nor on aptitude, as in other games, but on purely mathematical factors that predetermine it unconditionally.

Rubik's Cube

Rubik's cube was invented by the Hungarian Ernö Rubik in 1974, patented in 1975, and placed on the Hungarian market in 1977. By 1982, 10 million cubes had been sold in Hungary, more than the population of the country. It is estimated that over 100 million have been sold worldwide.

The cube is a $3 \times 3 \times 3$ array of smaller cubes that are colored so that the six faces of the large cube have six distinct colors. The twenty-six external subcubes are internally hinged so that rotation is possible. The goal of the puzzle is to return the cube to a state in which each side has a single color after it has been randomized by repeated rotations. The nine cubes forming one face can be rotated through forty-five rotations. There are

$$43,252,003,274,489,856,000$$

different arrangements of the small cubes, only one of these arrangements being the initial position where all colors match on each of the six sides. If you had a cube for every one of these "legal" positions, then you could cover the entire surface of the earth (including oceans) about 250 times. A column consisting of all the cube positions would stretch about 250 light-years.

Algorithms exist for solving a cube from an arbitrary initial position, but they are not necessarily optimal (i.e., requiring a minimum number of turns). The minimum number required for an arbitrary starting position is still unknown. In 1995 Michael Reid proved that the minimum number is less than or equal to twenty-nine turns (or forty-two "quarter-turns"). The proof involves large tables generated by computer.[35] Computers can be used to solve numerous random configurations. For example, Dik Winter wrote a program that has solved millions of random cubes in at most twenty-one turns.

These days, it's apparently an insufficient challenge to return a scrambled cube to a perfect state, so some have tried to produce elegant target patterns of various types. These patterns go by colorful names such as "spirals," "cherries," and "Christmas cross."[36] The Christmas cross is schematically represented in the following diagram where R = red, G = green, B = blue, Y = yellow, W = white, and O = orange.

R	W	R
W	W	W
R	W	R

W	R	W	G	Y	G	B	O	B	Y	G	Y
R	R	R	Y	Y	Y	O	O	O	G	G	G
W	R	W	G	Y	G	B	O	B	Y	G	Y

O	B	O
B	B	B
O	B	O

Christmas cross

In color, this looks beautiful. Why not invent your own variant of Rubik's cube and try to patent it? You can become the next Ernö

Rubik. The field is ripe; various patents have been granted for a range of puzzles closely related to Rubik's cube:

1. US3655201: "Pattern forming puzzle and method with pieces rotatable in groups," Larry Nichols, Moleculon Research Corporation, Cambridge, Massachusetts, 1972
2. US4421311: "Puzzle-cube," Peter Sebesteny, Ideal Toy Corporation, Hollis, New York, 1983
3. US5642884: "Holographic image reconstruction puzzle," David Pitcher, Polaroid Corporation, Cambridge, Massachusetts, 1997
4. USD0366506: "Game," Johan Lindquist, Sweden, 1995
5. USD0353850: "Three-dimensional puzzle," Karel Hrsel and Vojtech Kopsky, Czechoslovakia, 1993
6. US4872682: "Cube puzzle with moving faces," Ravi Kuchimanchi and Madhukar Thakur, Maryland and California, 1987
7. US4540177: "Puzzle cube," Tibor Horvath, New York, 1985
8. US4461480: "Educational entertainment device comprising cubes formed of four 1/8th octahedron sections rotatably coupled to a tetrahedron," Maurice Mitchell, California, 1984
9. US4067580: "Mystic numbered geometrics," John Tzeng, Maryland, 1976
10. US4258479: "Tetrahedron blocks capable of assembly into cubes and pyramids," Patricia Roane, California, 1981
11. US4513970: "Polymorphic twist puzzle," Ovidiu Opresco and Jon Marinesco, New York, 1985
12. US4605231: "Light transmission puzzle game," Lawrence Richman, Florida, 1986

Minh Thai of Vietnam holds the official (well documented) world record for solving a Rubik's cube from a scrambled state: 22.95 seconds. However, there are examples of other people who seem to have recorded faster times than this.[37] The mathematician John Conway is said to be able to solve the cube with only four or five glances at the cube; in other words, he performs most of the moves mentally, without looking at the cube.[38] There are many additional references regarding Rubik's cube.[39]

Rubik's Tesseract

Many readers are familiar with Ernö Rubik's ingenious cubical puzzle and its variations, which include a $4 \times 4 \times 4$ cube and puzzles shaped like a tetrahedra. One natural variation that never appeared on toy store shelves is the four-dimensional version of Rubik's cube–Rubik's tesseract. Dan Velleman of Amherst College discusses the $3 \times 3 \times 3 \times 3$ Rubik's tesseract in the February 1992 issue of *Mathematics Magazine*. Many of his findings were discovered with the aid of a colorful simulation on a Macintosh computer. Velleman remarks, "Of course, the tesseract is somewhat harder to work with than the cube, since we can't build a physical model and experiment with it." Those of you interested in pursuing the details of this mind-shattering tesseract should consult his paper.[40]

Here is a puzzle involving Rubik's tesseract that I posed in my book *Surfing Through Hyperspace*. It's based on Velleman's analysis.

Aliens have descended to Earth and placed a $3 \times 3 \times 3 \times 3$ foot Rubik's tesseract at the FBI Headquarters in Washington, D.C. (A tesseract is a four-dimensional cube in the same way that a cube is a three-dimensional version of a square.) The colors of this four-dimensional Rubik's cube shift every second for several minutes as onlookers stare and scream. (At first the FBI believes it to be a Russian spy device.) Finally, the tesseract is still–permitting us to scramble it by twisting any of its eight cubical "faces," as described in the following. You are sent to investigate.

You soon realize that this is an alien test, and humans have a year to unscramble the figure, or Washington, D.C., will be annihilated. Your question is: What is the total number of positions of the tesseract? Is the number greater or less than a trillion?

For a solution, see the end of this section.

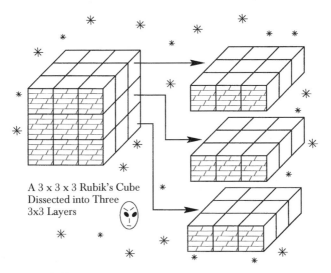

A 3 x 3 x 3 Rubik's Cube
Dissected into Three
3x3 Layers

Figure 2.23
Rubik's cube.

Here is some background to a four-dimensional Rubik's cube. Take a look at Figure 2.23, depicting a $3 \times 3 \times 3$ cubical puzzle. Each face is a 3×3 arrangement of small cubes called "cubies." If you were to cut this cube into three layers, each layer would look like a 3×3 square with the same four colors appearing along its sides. Two additional colors are in the interiors of all the squares in the first and third layers. (These are the colors on the bottom of all the squares in the first layer and the top of all the squares in the third layer.)

The aliens have extended this puzzle to the fourth dimension, where the four-dimensional $3 \times 3 \times 3 \times 3$ Rubik's hypercube, or tesseract, is composed of $3 \times 3 \times 3$ cubes stacked up in the fourth dimension. All cubes have the same six colors assigned to their faces, and in addition, there are two more colors assigned to the interiors of all of the little cubes (cubies) in the first and third cubes. (I refer to the eighty-one small cubes in this representation as "cubies," as have other researchers, such as Dan Velleman, although each is really one of the eighty-one small tesseracts that make up the alien Rubik's tesseract.)

The three- and four-dimensional puzzles differ in the following ways. The original Rubik's cube has six square faces. Rubik's tesseract has eight cubical "faces." In the standard Rubik's cube, there are three kinds of cubies: edge cubies with two colors, corner cubies with three colors, and face-center cubies with one color. (I ignore the cubie in the center of the cube, which has no color and plays no role in the puzzle.) Rubik's tesseract has four kinds of pieces that are also distinguished by the number of their colors.

Those of you with computers may enjoy "MagicCube4D," a fully functional four-dimensional analogue of Rubik's cube developed by Daniel Green and Don Hatch. The current implementation is for the Windows operating system. The graphical model is an exact 4-D extension of the original, plastic, three-dimensional puzzle, but with some useful features such as a "reset" button. Using the same mathematical techniques that are used to project 3-D objects onto 2-D screens, MagicCube software "projects" the 4-D cube into three dimensions. The resulting 3-D objects can then be rendered with conventional graphics software onto the screen. It is very difficult to solve the 4-D Rubik's cube starting from a scrambled initial structure. If you ever do succeed, you will be one of a *very* elite group of people. You will almost certainly need to have mastered the original Rubik's cube before you can hope to solve this one. Luckily, all of the skills learned for the original puzzle will help you solve the 4-D version. Also, you don't need to ever solve the full puzzle to enjoy it. For example, one fun game is to start with a slightly scrambled configuration, just a step or two away from the solved state, and work to back out those few random twists. If you get tired trying to solve the puzzle yourself, it is breathtaking to watch the computer solve the puzzle. For more information, visit http://www.superliminal.com/cube.htm.

Okay, here is the answer you have all been waiting for—the solution to the Rubik's tesseract question. The total number of positions of Rubik's tesseract is 1.76×10^{120}, far greater than a billion! As we saw in the last section, the total number of positions of Rubik's cube is 4.32×10^{19}. If either the cube or the tesseract changed positions every second since the beginning of the universe, they would still be turning today and would not have exhibited every possible configuration.

Addition-Multiplication Magic Square

Addition-multiplication magic squares are simultaneously magic squares and multiplication magic squares. For example, the eighth-order square shown here has an addition magic constant of 840 and a multiplicative magic constant of 2,058,068,231,856,000.[41]

46	81	117	102	15	76	200	203
19	60	232	175	54	69	153	78
216	161	17	52	171	90	58	75
135	114	50	87	184	189	13	68
150	261	45	38	91	136	92	27
119	104	108	23	174	225	57	30
116	25	133	120	51	26	162	207
39	34	138	243	100	29	105	152

Addition-multiplication magic square

Alphamagic Square

An *alphamagic square* is a magic square for which the number of letters in the word for each number generates another magic square.[42] The results depend, of course, on the language being used. In English, for example, we have

5	22	18
28	15	2
12	8	25

4	9	8
11	7	3
6	5	10

Alphamagic square   Resulting square

where the magic square on the right corresponds to the number of letters in

five	twenty-two	eighteen
twenty-eight	fifteen	two
twelve	eight	twenty-five

Numbers spelled out

In other words, you spell out the numbers in the first magic square and then count the letters in the words. The integers make a second magic square. This second square contains the consecutive digits from 3 to 11. This first square is referred to as an alphamagic square and was invented by puzzle enthusiast Lee Sallows, who made a thorough investigation of this type of square in 1986 and reported the results in *Abacus*. The square has since appeared in many publications.

Sign of the Beast

Add 100 to each cell of the previous two magic squares. Next, add the corresponding cells together to make a new magic square. The constant of this new square is 666.

Runes

The deciphering of the runic inscription below was what led Lee Sallows to the discovery of this new class of magic squares.

ᚠᛁᚠᛖ:ᛏᚹᛖᛏᛁᛋᛏᚹᚠ:ᚤᚻᛏᚠᛏᚾᛁᛖ

ᛏᚹᛖᛏᛁᛋᛗᚠᚻᛏᛖ:ᚠᛁᚠᛏᚾᛁᛖ:ᛏᚹᚠ

ᛏᚾᛖᛗᚸᚠ:ᛗᚠᚻᛏᛖ:ᛏᚹᛖᛏᛁᛋᚠᛁᚠᛖ

When translated, the inscription consists of a set of numbers that could be used to fill the spaces of a 3 × 3 square—which turned out to be the alphamagic square (the left array of numbers). As we've just discussed, remarkably, when the number in each space is replaced by the number of letters in the word for the number, a new magic square is created with numbers 3 to 11. It works in both the original language and in modern English.

As discussed in Ivars Petersen's *Islands of Truth*, Sallows's discovery led to a search for other examples in several languages that might have the same property. For columns, rows, and diagonals totaling less than 200, French has only one such magic square, whereas English has more than seven. Welsh, on the other hand, has more than twenty-six. For totals less than 100, none occurs in Danish, but six occur in Dutch, thirteen in Finnish, and an amazing 221 in German. There's even rumored to be a 3×3 English square from which a magic square can be derived, which in turn yields a third magic square. I do not know if an exhaustive search has been finished for 4×4 and 5×5 language-dependent magic squares. Sallows describes these kinds of quests as "searches for ever more potent magic spells."[43]

Bimagic Square

If replacing each number in a magic square by its square produces another magic square, the magic square is said to be a *bimagic square*.[44] The first discovered bimagic square (shown on the next page) is order 8 with magic constant 260 for addition and 11,180 after squaring. Bimagic squares are also called *doubly magic* squares or 2-multimagic squares. (See Gallery 1 for the first odd-ordered bimagic square ever discovered.)

Trimagic Square

If you can replace each number in a magic square by its square or cube and produce another magic square, the square is said to be a *trimagic square*.[45] Trimagic squares of order 32, 64, 81, and 128 are known. There are methods available for constructing a trimagic square of order 64, 81, or 128. Trimagic squares are also called *trebly magic squares* and 3-multimagic squares.

16	41	36	5	27	62	55	18
26	63	54	19	13	44	33	8
1	40	45	12	22	51	58	31
23	50	59	30	4	37	48	9
38	3	10	47	49	24	29	60
52	21	32	57	39	2	11	46
43	14	7	34	64	25	20	53
61	28	17	56	42	16	6	35

Eighth-order bimagic square

Multimagic Square

A magic square is *p-multimagic* if the square formed by replacing each element by its kth power (for $k = 1, 2, \ldots, p$) is also magic.[46] As we discussed, a 2-multimagic square is called a bimagic square, and a 3-multimagic square is called a trimagic square. I do not know if a quadramagic or pentamagic square exists and welcome feedback from readers.

Latin Square

A Latin square consists of n sets of numbers 1 to n arranged in such a way that no orthogonal (row or column) contains the same two numbers.[47] The numbers of Latin squares of order $n = 1, 2, 3, 4, \ldots$ are 1, 2, 12, 576, 161280, 812851200, 61479419904000, 108776032459082956800, 5524751496156892842253122560, 9982437658213039871725064756920320000, ...

A pair of Latin squares is said to be orthogonal if the n^2 pairs formed by juxtaposing the two arrays are all distinct. ("Juxtaposed" means combining the two numbers to form an ordered pair.)

The two Latin squares of order 2 are given by

1	2
2	1

2	1
1	2

Latin squares of order 2

Two orthogonal Latin squares of order 3 are

3	2	1
2	1	3
1	3	2

2	3	1
1	2	3
3	1	2

Order-3 Latin square   Order-3 Latin square

Juxtaposing the two squares yields the following n^2 pairs:

$$(3,2)\ (2,3)\ (1,1)$$
$$(2,1)\ (1,2)\ (3,3)$$
$$(1,3)\ (3,1)\ (2,2)$$

Two of the 576 Latin squares of order 4 are

1	2	3	4
2	1	4	3
3	4	1	2
4	3	2	1

1	2	3	4
3	4	1	2
4	3	1	2
2	1	4	3

Order-4 Latin square   Order-4 Latin square

Gnomon Magic Square

A *gnomon magic square* is an $N \times N$ array of numbers in which the elements in each $N - 1 \times N - 1$ corner have the same sum.[48] For example, a 3×3 array can be cut up into a 2×2 array with a five-element L-shaped piece in four ways. The L-shaped piece is called a *gnomon*. When the sums of the elements in the four 2×2 arrays are equal, the larger array is said to be a gnomon magic square. For example, in the square array

1	8	3
6	5	4
7	2	9

Gnomon magic square

the sum of the elements in each 2×2 array is the magic constant 20. One gnomon is highlighted. Similarly, if the four sums are all different, the array is a *gnomon-antimagic square.*[49]

Magic Labeling

It is conjectured that every tree (branching structure) with e edges whose nodes are all trivalent or monovalent can be given a "magic" labeling such that the integers $1, 2, \ldots , e$ can be assigned to the edges so that the sum of the three meeting at a node is constant.[50] A tree is a mathematical structure that can be viewed as a graph with a set of straight-line segments connected at their ends containing no closed loops (cycles). A tree with n nodes has $n-1$ edges. The points of connection are known as forks and the segments as branches. Terminal segments and the nodes at their ends are called *leaves*. A tree with two branches at each fork and with one or two leaves at the end of each branch is called a binary tree. A trivalent tree is a tree in which every vertex, except the root and endpoints, has three edges connecting to it. Figure 2.24 shows a magically labeled trivalent tree in which all the edges surrounding a trivalent node sum to 18.

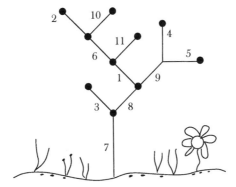

Figure 2.24
Tree with magic labeling.

Magic Hexagon (Hexagonal Tiling)

A *magic hexagon* or hexagonal tiling is an arrangement of close-packed hexagons containing the numbers $1, 2, \ldots, H_n = 3n(n-1) + 1$, where H_n is the nth hexagonal number, such that the numbers along each straight line add up to the same sum. Gallery 3 exhibits a magic hexagon, each of whose lines (those of lengths 3 and 4) adds up to 38. This is the only magic hexagon of the positive integers for any size hexagon, as proved by Charles Trigg.[51] The magic hexagon was discovered by C. W. Adams, who worked on the problem from 1910 to 1957. Charles Trigg showed that the magic constant for an order n hexagon would be

$$\frac{9(n^4 - 2n^3 + 2n^2 - n) + 2}{2(2n - 1)}$$

which requires $5/(2n-1)$ to be an integer in order for a solution to exist. But this is an integer for only $n = 1$ (the trivial case of a single hexagon), and Adams's $n = 3$.[52]

Magic Interlocked Hexagons

Hexagons may be packed tightly side by side on the surface of a torus and each hexagon divided into six identical triangles (Fig. 2.25). *Magic interlocked hexagons* are figures in which the six triangles

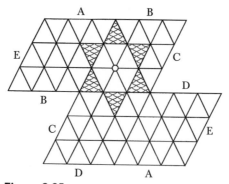

Figure 2.25
Interlocked hexagons can tile a torus if you connect the corresponding edges. Your mission is to fill each triangle so that the six triangles around every vertex contain numbers that sum to 219. The six triangles adjacent to these (forming a starlike shape) also must have this sum.

around every vertex (an example is shown as a circle) contain numbers that sum to 219. The six triangles adjacent to these (forming a starlike shape) also have this sum. A specific example with numbers is exhibited in Gallery 3.

Multiplication Magic Square

A *multiplication magic square* is a square that is magic under multiplication instead of the traditional operation of addition.[53] Unlike standard magic squares, the N^2 entries for an Nth-order multiplicative magic square are not required to be consecutive. The multiplication magic square here has a multiplicative magic constant of 4096.

128	1	32
4	16	64
8	256	2

Multiplication magic square

Can you see how this square might be constructed from an additive magic square?

Magic Series

A *magic series*[54] of degree p is formed by n numbers if the sum of their kth powers is the magic constant of degree k for every k from 1 to p. Sound complicated? Note 54 gives additional information.

Domino Magic Squares

Dominoes is a game played with small, rectangular pieces identified by a number of dots, or pips, on their faces. The face of each piece is divided by a line into two squares, each of which is marked as would be a pair of dice, except that some squares are blank (indicating zero). The usual set consists of twenty-eight pieces, marked respectively: 6-6 ("double six"), 6-5, 6-4, 6-3, 6-2, 6-1, 6-0, 5-5, 5-4, 5-3, 5-2, 5-1, 5-0, 4-4, 4-3, 4-2, 4-1, 4-0, 3-3, 3-2, 3-1, 3-0, 2-2, 2-1, 2-0, 1-1, 1-0, 0-0.

In China, dominoes were used in the twelfth century A.D. They seem to have been designed to represent all possible throws with two dice, for Chinese dominoes have no blank faces. Western dominoes were probably not derived from the Chinese. There is no record of them before the mid-eighteenth century in Italy and France. Apparently they were introduced to England by French prisoners toward the end of the eighteenth century. The North American Eskimos also play a dominolike game, using sets consisting of as many as 148 pieces.

Domino magic squares contain an arrangement of dominoes in which the fifty-six squares form a 7×7 magic square. The sum of the pips in each row, column, and diagonal of the square are equal. Examples are given in Gallery 3.

Supermagic and Antimagic Graphs

A graph with q edges is called *supermagic* if it is possible to label the edges with the numbers 1, 2, 3, ..., q in such a way that at each vertex v, the sum of the labels on the edges incident with v is the same. Extensive work in this area has been conducted by Nora

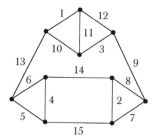

Figure 2.26
Supermagic graph.

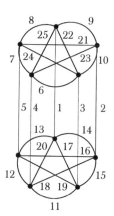

Figure 2.27
Supermagic graph.

Hartsfield of the Western Washington University in Bellingham and Gerhard Ringel of the University of California, Santa Cruz.[55]

Figure 2.26 shows an example in which the magic constant is 24. For example, $10 + 1 + 13 = 24$, $11 + 12 + 1 = 24$, etc. Figure 2.27 shows a three-dimensional supermagic graph. Notice how all the edge labels around a vertex give the same magic constant; for example, the node at top and center yields $8 + 25 + 22 + 9 + 1 = 65$, the node at upper right yields $9 + 21 + 23 + 10 + 2 = 56$, and so forth.

Figure 2.28 shows examples of two graphs whose edges are labeled with the integers $1, 2, 3, \ldots, q$, so that the sum of the labels at any given vertex is *different* from the sum of the labels at any other vertex; that is, no two vertices have the same sum. These

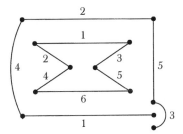

Figure 2.28
Antimagic graph.

graphics are called *antimagic*. I would be interested to hear from readers who have constructed supermagic and antimagic graphics in higher dimensions.

Annihilation Magic Squares

Annihilation magic squares are those in which the rows, columns, and two main diagonals sum to zero.[56] For these squares, consecutive numbers from $-N/2$ to $N/2$ are required, and zero is excluded. Below is an example:

8	−7	−6	5
−4	3	2	−1
1	−2	−3	4
−5	6	7	−8

Annihilation magic square

Can you find others of this type? Can you find a 3×3 magic square such that by subtracting the central number–in any row, column, or diagonal–from the sum of the other two numbers, the result is the same?

Crystal Magic Squares

Crystal magic squares[57] have many similarities to magic squares but have a slightly different geometry, as illustrated here. Boris

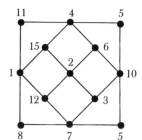

Figure 2.29
Crystal magic square.

Kordemsky, author of *The Moscow Puzzles,* writes

> We have a crystal lattice whose "atoms" are joined in ten rows
> of three atoms each. Select thirteen integers, of which twelve
> are different, and place them in the "atoms" so that each row
> totals 20. (The smallest number needed is 1, the largest 15.)

Figure 2.29 is one such solution. Are there others? Can you extend
this to larger crystal lattices or to cubical lattices? Can you find a
solution with consecutive integers that are all different?

Magic Word Squares

Magic word squares have been thoroughly researched by Jeremiah
Farrell of Indianapolis, Indiana (for example, see his paper
"Magic Square Magic: World Ways" in the May 2000 issue of *The
Journal of Recreational Linguistics,* 33(2):83–92). Each cell of the
square must contain a dictionary entry from the *American Heritage
Dictionary* or *Merriam-Webster's New International Dictionary.* No letter is duplicated in any row or column. The main diagonals also
have this property for the fourth- and fifth-order magic word
squares. Additionally, every row and column must contain the
letters of the name of the magic square. Finally, each square has a
mathematical magic square counterpart when numbers are
assigned to letters. The following examples should clarify the
concept.

AD	IN	SO
IS	DO	AN
NO	AS	ID

ADONIS

AS	IR	ED	TO
DO	ET	IS	RA
IT	AD	OR	ES
RE	SO	AT	ID

ASTEROID

EN	MA	IR	SO	UT
IS	TO	NU	ME	RA
MU	RE	AS	IT	NO
AT	IN	OM	UR	ES
OR	US	ET	AN	MI

MOUSTERIAN

Believe it or not, each of the two-letter fragments is an entry in the English dictionary. For example, *et* means small one, *ut* is a musical tone, and *mu* is a letter in the Greek alphabet. Each of the squares has a mathematical counterpart. For ADONIS, set $A = 0$, $O = 1$, $I = 2$, $N = 0$, $D = 3$, and $S = 6$, and then add the numerical values of each of the letter pairs. Doing this, one creates a magic square with integers 0 through 8. Similarly, ASTEROID becomes numerically magic with the following assignments: $A = 0$, $E = 1$, $I = 2$, $O = 3$, $S = 0$, $T = 4$, $R = 8$, and $D = 12$. For MOUSTERIAN, set $A = 0$, $E = 1$, $I = 2$, $O = 3$, $U = 4$, $M = 0$, $N = 5$, $R = 10$, $S = 15$, and $T = 20$. (For word scholars among you, the definition of *mouseterian* is "of or relating to a lower Paleolithic culture.") In each case, the numerical magic constant is the number obtained by adding all letter values in the word constant.

Jeremiah Farrell uses the same procedure to create *magic word cubes*. In the MOUSETRAP cube, every set of three words, in any of the three dimensions, transposes into MOUSETRAP. The cube can be turned into a numerical one by determining the "misgraph" of the configuration. The misgraph is the graph of letters that are never used together to form any of the words in the cells. Here the misgraph is the disjoint set OAU (Hawaiian: "to mew as a cat"), MRS, and PET. Set $O = 0$, $A = 1$, $U = 2$, $M = 0$, $R = 3$, $S = 6$, $P = 0$, $E = 9$, and $T = 18$ to obtain a number magic cube.

MOP	RUE	SAT
RAT	SOP	EMU
USE	MAT	PRO

EAR	SOT	UMP
SUP	MAE	ROT
TOM	PUR	SEA

STU	MAP	ORE
MOE	RUT	SAP
RAP	OSE	TUM

Top layer   Middle layer   Bottom layer

The MOUSETRAP cube

These methods may be extended to magic word stars and used to create clever magic tricks. No magic word tesseract has ever been created.

Magic Rectangles

In the July 2000 issue of *Scientific American,* Ian Stewart discusses *magic rectangles*, $M \times N$ arrays of the integers ranging from 1 to the product of M and N. The numbers in each row add up to the same sum S_1, and the numbers in each column add up to the same sum S_2, but S_1 need not equal S_2. The diagonals are not considered.[58] Magic rectangles exist when M and N are either both even or both odd, provided that M and N are greater than 1 and are not both equal to 2. Thomas R. Hagedorn of the College of New Jersey generalizes this idea to higher-dimensional constructs.[59] For example, if all sides of a higher-dimensional array of integers are even (for example, a $2 \times 4 \times 8 \times 20$ "rectangle"), then a magic "rectangle" must exist.

Here are some example magic rectangles:

1	7	6	4
8	2	3	5

1	11	3	9	8	7
12	2	10	4	5	6

1	15	3	13	12	6	10	8
16	2	14	4	5	11	7	9

6	7	8	9	10
13	3	1	11	12
5	14	15	4	2

8	9	10	11	12	13	14
4	5	16	2	15	17	18
21	19	7	20	6	3	1

CHAPTER THREE

Gallery 1: Squares, Cubes, and Tesseracts

*Although some of the greatest mathematicians have done work on
magic squares, and even though such work leads into the theories of
groups, lattices, Latin squares, determinants, partitions, matrices,
congruence arithmetic and other nontrivial areas of mathematics, the
most enthusiastic square constructors have been amateurs.*

–Martin Gardner, *Time Travel and Other Mathematical
Bewilderments*

This chapter contains an exhibition gallery of my favorite magic
squares and related constructs such as magic cubes and tesseracts.
The objects span the centuries, with some arrangements so exotic
that they seem to defy easy classification. I suspect that many contain hidden patterns yet to be articulated. Let's start with a beautiful square from the eighteenth century.

Ben Franklin's "Most Magically Magical" Square

Excluding George Washington, Benjamin Franklin (1706–1790)
was the most famous eighteenth-century American. He was a scientist, inventor, statesman, printer, philosopher, musician, and
economist. It is easy to see how such a curious person could create
one of the most fascinating squares ever conceived. After Franklin's death, other magic square researchers have discovered new
patterns in Franklin's number arrays. That's the breathtaking
aspect of magic squares. Armed with just a pencil and paper, you
can discover new patterns in centuries-old magic squares. Each
square is like a little treasure box waiting to be opened.

Let me have Ben Franklin introduce his wondrous square in his own words (from "Letters and papers on Philosophical subjects by Benjamin Franklin, L.L.D., F.R.S.," printed in London in 1769).[1] Ben Franklin starts by wondering if magic squares are trivial time wasters and then proceeds to shatter our minds with incredible squares of his own invention.

From: Benjamin Franklin Esq. of Philadelphia

To: Peter Collinson Esq. At London

According to your request I now send you the arithmetical curiosity of which this is the history.

Being one day in the country at the house of our common friend, the late learned Mr. Logan, he showed me a folio French book filled with magic squares, wrote, if I forget not, by one M. Frénicle [Bernard Frénicle de Bessy], in which, he said, the author had discovered great ingenuity and dexterity in the management of numbers; and, though several other foreigners had distinguished themselves in the same way, he did not recollect that any one Englishman had done anything of the kind remarkable.

I said it was perhaps a mark of the good sense of our English mathematicians that they would not spend their time in things that were merely *difficiles nugae*, incapable of any useful application. Logan answered that many of the mathematical questions publicly proposed in England were equally trifling and useless. Perhaps the considering and answering such questions, I replied, may not be altogether useless if it produced by practice an habitual readiness and exactness in mathematical disquisitions, which readiness may, on many occasions be of real use. In the same way, says he, may the making of these squares be of use.

I then confessed to him that in my younger days, having once some leisure which I still think I might have employed more usefully, I had amused myself in making these kind of magic squares and, at length, had acquired such a knack at it, that I could fill the cells of any magic square, of reasonable size, with a series of numbers as fast as I could write them, disposed in such a manner, as that the sums of every row, horizontal, perpendicular, or diagonal, should be equal; but not being satisfied with these, which I looked on as common and easy things, I had imposed on myself more difficult tasks, and succeeded in making other magic squares, with a variety of properties, and much more curious. He then showed me several in the same book, of an uncommon and more curious kind; but, as I thought none of them equal to some I remembered to have made, he desired me to let him see them; and accordingly, the next time I visited him, I carried him a square of 8, which I found among my old papers and which I now give you, with an account of its properties.

52	61	4	13	20	29	36	45
14	3	62	51	46	35	30	19
53	60	5	12	21	28	37	44
11	6	59	54	43	38	27	22
55	58	7	10	23	26	39	42
9	8	57	56	41	40	25	24
50	63	2	15	18	31	34	47
16	1	64	49	48	33	32	17

Eighth-order Franklin square

The properties are:

1. That every straight row (horizontal or vertical) of 8 numbers added together, makes 260, and half of each row half of 260.

2. That the bent row of 8 numbers, ascending and descending diagonally, viz. from 16 ascending to 10, and from 23 descending to 17; and every one of its parallel bent rows of 8 numbers make 260, etc. [See gray highlighted squares for two examples of "bent rows." See thickly highlighted squares for an example "broken bent row" (14 + 61 + 64 + 15 + 18 + 33 + 36 + 19), which also sums to 260.]

3. And, lastly, the 4 corner numbers with the 4 middle numbers make 260.

So this magical square seems perfect in its kind. But these are not all its properties; there are 5 other curious ones, which, at some other time I will explain to you.

Mr. Logan then showed me an old arithmetical book in quarto, wrote, I think, by one [Michel] Stifelius, which contained a square of 16×16 that he said he should imagine must have been a work of great labor; but I forget not, it had only the common properties of making the same sum, viz., 2056, in every row, horizontal, vertical, and diagonal. Not willing to be outdone by Mr. Stifelius, even in the size of my square, I went home and made that evening the following magical square of 16, which, besides having all the [special] properties of the 8×8 square (i.e. it would make 2056 in all the same rows and bent and broken bent rows), had this added: that a four-square hole being cut in a piece of paper of such a size as to take in and show through it just 16 of the little squares, when laid on the greater square, the sum of the 16 numbers so appearing through the hole, wherever it was placed on the greater square, should likewise make 2056.

This I sent to our friend the next morning, who, after some days, sent it back in a letter with these words:

"I return to thee thy astonishing
or most stupendous piece
of the magical square, in which . . ."

–but the compliment is too extravagant, and therefore, for his sake as well as my own, I ought not to repeat it. Nor is it

200	217	232	249	8	25	40	57	72	89	104	121	136	153	168	185
58	39	26	7	250	231	218	199	186	167	154	135	122	103	90	71
198	219	230	251	6	27	38	59	70	91	102	123	134	155	166	187
60	37	28	5	252	229	220	197	188	165	156	133	124	101	92	69
201	216	233	248	9	24	41	56	73	88	105	120	137	152	169	184
55	42	23	10	247	234	215	202	183	170	151	138	119	106	87	74
203	214	235	246	11	22	43	54	75	86	107	118	139	150	171	182
53	44	21	12	245	236	213	204	181	172	149	140	117	108	85	76
205	212	237	244	13	20	45	52	77	84	109	116	141	148	173	180
51	46	19	14	243	238	211	206	179	174	147	142	115	110	83	78
207	210	239	242	15	18	47	50	79	82	111	114	143	146	175	178
49	48	17	16	241	240	209	208	177	176	145	144	113	112	81	80
196	221	228	253	4	29	36	61	68	93	100	125	132	157	164	189
62	35	30	3	254	227	222	195	190	163	158	131	126	99	94	67
194	223	226	255	2	31	34	63	66	95	98	127	130	159	162	191
64	33	32	1	256	225	224	193	192	161	160	129	128	97	96	65

The most magically magical of any magic square *ever* made

necessary; for I make no question but you will readily allow this square of 16 to be *the most magically magical* of any magic square *ever* made by any magician.

I am etc. B. F.

Wasn't Franklin a remarkable man? In his letter, he described an 8 × 8 magic square he had devised in his youth, and the special properties it possessed. The square is indeed filled with wondrous symmetries, some of which Ben Franklin was probably not aware.

As explained by Franklin, each row and column of the square have the common sum 260. Also, he noted that half of each row or column sums to half of 260. In addition, each of the "bent rows" (as Franklin called them) have the sum 260. Here is a schematic representation of some of the patterns that produce magic sums.

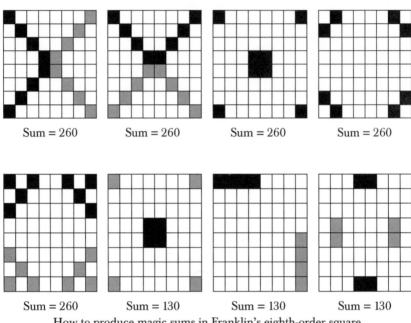

Sum = 260   Sum = 260   Sum = 260   Sum = 260

Sum = 260   Sum = 130   Sum = 130   Sum = 130

How to produce magic sums in Franklin's eighth-order square

Every time I look at this beauty, I find more symmetries and patterns. The sum of the numbers in any 2×2 subsquare is 130, and the sum of any four numbers that are arranged equidistant from the center of the square also equals 130.

Some of the numerical patterns can be shifted up, down, right, and left (with wrap-around as if the 8×8 pattern were on the surface of a doughnut with opposite ends joined) still producing the 260 sum! What other patterns can you find?

Notice that other symmetries are made obvious if we subtract 1 from each number and convert the numbers to binary, or base 2, numbers:[2]

110011	111100	000011	001100	010011	011100	100011	101100
001101	000010	111101	110010	101101	100010	011101	010010
110100	111011	000100	001011	010100	011011	100100	101011
001010	000101	111010	110101	101010	100101	011010	010101
110110	111001	000110	001001	010110	011001	100110	101001
001000	000111	111000	110111	101000	100111	011000	010111
110001	111110	000001	001110	010001	011110	100001	101110
001111	000000	111111	110000	101111	100000	011111	010000
1	2	3	4	5	6	7	8

Ben Franklin disintegrates in an orgy of self-annihilation

(Binary numbers are defined in note 3.) Pretend for a moment that the 1s represent matter and the 0s represent antimatter. If we superimpose column 1 on 4, 2 on 3, 5 on 8, and 6 on 7, we find that the bit patterns are exact opposites: 110011 annihilates 001100 and so forth, until the entire square self-destructs in an orgy of annihilation.

In order to hunt for patterns and symmetries, I also like to shade magic squares so that even cells go from white to black and odd cells go from black to white. Here is Ben Franklin's order-8 square colored in such a manner. Perhaps you can think of more interesting ways of shading each cell in order to reveal patterns.

52	61	4	13	20	29	36	45
14	3	62	51	46	35	30	19
53	60	5	12	21	28	37	44
11	6	59	54	43	38	27	22
55	58	7	10	23	26	39	42
9	8	57	56	41	40	25	24
50	63	2	15	18	31	34	47
16	1	64	49	48	33	32	17

Ben Franklin tiled

Franklin's 16×16 square is even more awesome than his eighth-order square, and it is no exaggeration to say that one could spend a lifetime contemplating its wonderful structure. In this square Franklin discovered that the sum of the numbers in any 4×4 sub-square is 2056. The fun has only started. Notice that the sum of the numbers in any 2×2 subsquare is 514. Notice that the various sums can be formed by simple expressions of powers of 2: $2056 = 2^{11} + 2^3$ and $514 = 2^9 + 2^1$. Additionally, the square has all kinds of bent row properties and symmetries similar to those of the 8×8 square. Try drawing some diagrams for the 16×16 square showing cell configurations that produce the magic sum.

I hate to be a party-pooper, but, sadly, despite all the marvelous symmetries, neither of Franklin's squares satisfies the main diagonal sums, so they cannot strictly qualify as "magic squares" according to the common definition that includes the diagonal sums. Did God throw a monkey wrench into the works to destroy some of the perfection and make humans wonder where they went wrong? Yet, despite these irregularities, there is still beauty. In the 16×16 square, one diagonal sums to $2056 - 128$ and the other sums to $2056 + 128$. The number 128 is a power of 2.

We do not know what method Franklin used to construct his squares. Many people have tried to crack the secret, but until the 1990s no *quick* recipe could be found, and Franklin claimed he could generate the squares "as fast as he could write." Magic square experts such as Jim Moran, author of *The Wonders of Magic Squares*, suggest that many people believed Ben Franklin was stretching the truth:

> Ben has always been one of my idols, and now he has gone and tarnished his image and revealed himself to me as being merely human. In telling his colleague Collinson that he could make his magic squares as fast as he could write down the numbers, he was pushing reasonable exaggeration beyond the limit. I just can't buy this, and I say nobody but *nobody* can perform this feat. He even claimed to do this with squares of any reasonable size. *No way!*[4]

In the same book, Moran showed a complicated method of constructing Franklin's 8×8 square, but he himself admits:

> We'll never know for sure how Dr. Franklin put together his fantastic 8×8 square. All we can say is "Bravo, Ben!" As he states in his letter to Peter Collinson, he had the knack of arranging the numbers in magic order as fast as he could write them down, and there is no reason to disbelieve him (or is there?). He doubtlessly used a method far superior to the rather laborious one presented here, but we are at least able to show one possible approach.

In 1991, Lalbhai D. Patel invented another method to construct the Franklin squares. Although the method seems quite long, Patel has trained himself to quickly carry out the procedure. You can read all the gory details of the approach in his *Journal of Recreational Mathematics* article.[5]

We've discussed a few ways to reveal patterns in magic squares, such as different shadings for odd and even numbers, representing the numbers as binary numbers, highlighting broken diagonals, and so forth. However, even when armed with these techniques, we know that Franklin's squares possess a beauty and symmetry that can be made more obvious if we make geometrical diagrams that list all of the numbers in the square and connect those that are in the same rows. Figure 3.1 is a geometrical diagram for the 8×8 square. Similar kinds of representations were perhaps first drawn for the Franklin squares around 1917,[6] and I like to meditate upon them as if they are mandalas–in Hindu and Buddhist Tantrism, the symbolic diagram used in the performance of sacred rites and as an instrument of meditation. Traditionally, mandalas were painted on paper or cloth, drawn on a carefully prepared ground with white and colored threads or with rice powders, fashioned in bronze, or built in stone. Perhaps mandalas of the twenty-first century will include the geometric diagrams for huge magic squares with dimensions greater than 1000×1000. Perhaps computers will automatically draw them instead of humans.

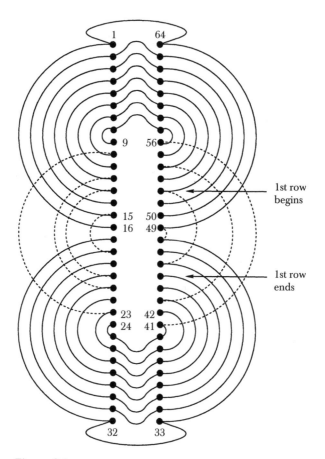

Figure 3.1

Geometrical diagram for Franklin's 8 × 8 square.

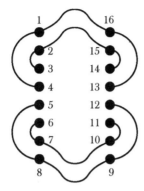

Figure 3.2
Geometrical diagram for a
4 × 4 magic square.

In addition to their intrinsic beauty, the diagrams have a practical side, especially when they give us hints about the existence of higher-dimensional magic figures. Consider the little mandala in Figure 3.2 that represents the following 4 × 4 magic square with magic bent diagonals:

5	8	9	12
14	15	2	3
11	10	7	6
4	1	16	13

Square with magic bent diagonals

The bent diagonals (one in gray) add up to 34. We can repeat the little mandala's theme in Figure 3.2 four times and add a symmetrical set of interconnections to form another geometrical diagram (Fig. 3.3) that corresponds to a magic cube consisting of four 4 × 4 layers. The following arrays are cross sections of the 4 × 4 × 4 cube; the upper left square and bottom right square are the top and bottom layers of the cube, respectively.

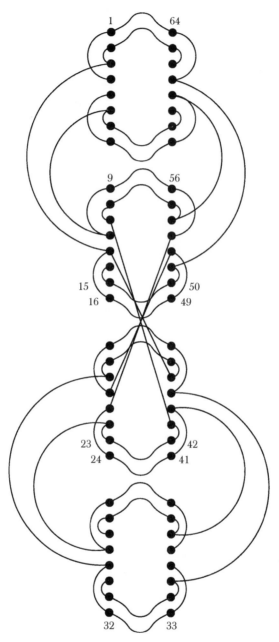

Figure 3.3

Geometrical diagram for a 4 × 4 × 4 cube.

5	8	57	60
54	55	10	11
43	42	23	22
28	25	40	37

59	58	7	6
12	9	56	53
21	24	41	44
38	39	26	27

62	63	2	3
13	16	49	52
20	17	48	45
35	34	31	30

4	1	64	61
51	50	15	14
46	47	18	19
29	32	33	36

Cross sections of a fourth-order magic cube

It would be interesting to draw such diagrams for large squares, cubes, and four-dimensional magic figures, and I would be interested in hearing from readers who attempt this feat.

Prime Number Bordered Square

On the next page is an awesome 13×13 magic square created by a man in jail[7] and composed of prime numbers. (Prime numbers such as 5, 7, 11, and 1153 are evenly divisible only by 1 and themselves.) The square is also a "bordered square"; the magic square contains various subsquares whose borders can be peeled away creating successive magic squares of size 11×11, 9×9, 7×7, 5×5, and 3×3. The magic constants for the subsquares are 70,681, 59,807, 48,933, 38,059, 27,185, and 16,311, respectively.

The common difference between these magic constants is 10,874, a sum applying even to the center integer 5437 and the magic constant of the 3×3 square. I suspect that there are many other patterns for you to find. For example, just by chance I added up the four corner squares $(1153 + 8353 + 9721 + 2521)$ and got $10,874 \times 2$, which is also four times 5437, the value of the center square. Isn't this beautiful? The same holds for the corners of the

1153	8923	1093	9127	1327	9277	1063	9133	9611	1693	991	8887	8353
9967	8161	3253	2857	6823	2143	4447	8821	8713	8317	3001	3271	907
1831	8167	4093	7561	3631	3457	7573	3907	7411	3967	7333	2707	9043
9907	7687	7237	6367	4597	4723	6577	4513	4831	6451	3637	3187	967
1723	7753	2347	4603	5527	4993	5641	6073	4951	6271	8527	3121	9151
9421	2293	6763	4663	4657	9007	1861	5443	6217	6211	4111	8581	1453
2011	2683	6871	6547	5227	1873	5437	9001	5647	4327	4003	8191	8863
9403	8761	3877	4783	5851	5431	9013	1867	5023	6091	6997	2113	1471
1531	2137	7177	6673	5923	5881	5233	4801	5347	4201	3697	8737	9343
9643	2251	7027	4423	6277	6151	4297	6361	6043	4507	3847	8623	1231
1783	2311	3541	3313	7243	7417	3301	6967	3463	6907	6781	8563	9091
9787	7603	7621	8017	4051	8731	6427	2053	2161	2557	7873	2713	1087
2521	1961	9781	1747	9547	1597	9811	1741	1213	9181	9883	1987	9721

Bordered square constructed while in prison

next layer of the onion: $7603 + 2713 + 8161 + 3271 = 10{,}874 \times 2$. Does this relation hold for all corners?

Binary Dürer's Square

Mark Collins is a colleague from Madison, Wisconsin, who has an interest in both number theory and artist Albrecht Dürer's works. Mark has studied the Dürer square discussed in the Introduction and finds some astonishing features when the numbers are converted to *binary code*. (In the binary representation, numbers are written in a positional number system that uses only two digits, 0 and 1—as explained in note 3 for this chapter.) Since the first sixteen hexadecimal binary numbers start with the number 0 and end with 15, he subtracts 1 from each entry in the magic square. The following is the result:

$$
\begin{array}{cccc}
15 & 2 & 1 & 12 \\
1111 & 0010 & 0001 & 1100 \\
8 & 5 & 6 & 11 \\
1000 & 0101 & 0110 & 1011 \\
3 & 14 & 13 & 0 \\
0011 & 1110 & 1101 & 0000
\end{array}
$$

Remarkably, if the binary representation for the magic square is rotated 45 degrees clockwise about its center so that the 15 is up and the 0 down, the resultant pattern has a vertical mirror plane down its center:

$$
\begin{array}{ccccccc}
& & & 1111 & & & \\
& & 0100 & & 0010 & & \\
& 1000 & & 1001 & & 0001 & \\
0011 & & 0101 & & 1010 & & 1100 \\
& 1110 & & 0110 & & 0111 & \\
& & 1101 & & 1011 & & \\
& & & 0000 & & &
\end{array}
$$

For example, in row 2, 0100 is the mirror of 0010. (I very much doubt that Dürer could have known about this symmetry.)

If we rotate the square counterclockwise so that the 12 is at the top and the 3 at the bottom, we get a pattern that has a peculiar left-right inverse when drawing an imaginary vertical mirror down the center of the pattern.

$$
\begin{array}{ccccccc}
& & & 1100 & & & \\
& & 0001 & & 0111 & & \\
& 0010 & & 1010 & & 1011 & \\
1111 & & 1001 & & 0110 & & 0000 \\
& 0100 & & 0101 & & 1101 & \\
& & 1000 & & 1110 & & \\
& & & 0011 & & &
\end{array}
$$

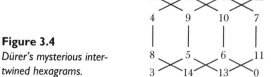

Figure 3.4
*Dürer's mysterious inter-
twined hexagrams.*

For example, in the second row, 0001 and 0111 are mirror inverses of each other.

Mark Collins has rediscovered the presence of mysterious intertwined hexagrams when the even and odd numbers are connected (Fig. 3.4; see also Fig. 8 in the Introduction).

I would be interested in hearing from those of you who find additional meaning or patterns in Dürer's magic square. What patterns can you find in other magic squares when numbers are converted to binary numbers? Mark has also done numerous experiments converting these numbers to colors, and he comments: "I believe this magic square is an archetype as rich in meaning and mysticism as the I-Ching. I believe it is a mathematical and visual representation of nature's origami—as beautiful as a crystal lattice or a photon of light." Mark suggests you should create other mitosis-like diagrams by connecting 0 to 1 to 2 to 3, then 4 to 5 to 6 to 7, then 8 to 9 to 10 to 11, and finally 12 to 13 to 14 to 15.

Apocalyptic Magic Square

A rather "beastly" 6 × 6 magic square was invented by the mysterious A. W. Johnson. No one knows when this square was constructed, nor is there much information about Johnson. (I welcome any information you may have.) All of its entries are prime numbers, and each row, column, diagonal, and broken diagonal sums to 666, the Number of the Beast. (As discussed, a broken diagonal is the diagonal produced by wrapping from one side of the square to the other; for example, the outlined numbers 131, 83, 199, 113, 13, and 127 form a broken diagonal.)

3	107	5	131	109	311
7	331	193	11	83	41
103	53	71	89	151	199
113	61	97	197	167	31
367	13	173	59	17	37
73	101	127	179	139	47

The apocalyptic magic square

Another Beastly Magic Square

The following order-6 magic square was constructed by Patrick De Geest from the first 36 multiples of 6, and has a magic sum of 666.[8] This square contains many sums that are three-digit palindromes. For example, the top left 3×3 square is magic with $\mathcal{S} = 252$. The bottom left 3×3 square is magic with $\mathcal{S} = 414$. The three rows of three cells in the top right corner sum to 414. The three rows of three cells in the bottom right corner sum to 252. The corners of the three squares formed working from the outside to the center each sum to 444. The 6×6 border cells sum to 2220, which equals $666 + 888 + 666$. The border cells of the central 4×4 square sum to 1332, which equals $666 + 666$. The top half of the right-hand column sums to 252, and the bottom half sums to 414. The top half of the column next to it sums to 414, and the bottom half sums to 252. By dividing each

66	108	78	174	216	24
96	84	72	204	30	180
90	60	102	198	168	48
120	162	132	12	54	186
150	138	126	42	192	18
144	114	156	36	6	210

Another beastly square

number in the magic square by 6, a new magic square is obtained with $S = 111$. What other amazing features can you find?

The Kurchan Square

Another amazing magic square is the *Kurchan square*, named after its discoverer, Rodolfo Marcelo Kurchan from Buenos Aires, Argentina.[9] He believes this to be the smallest, nontrivial magic square having N^2 distinct pandigital integers and the smallest pandigital magic sum. *Pandigital* means all ten digits are used, and zero is not the leading digit. Following is the awesome Kurchan Array; the pandigital sum is 4,129,607,358:

1,037,956,284	1,036,947,285	1,027,856,394	1,026,847,395
1,026,857,394	1,027,846,395	1,036,957,284	1,037,946,285
1,036,847,295	1,037,856,294	1,026,947,385	1,027,956,384
1,027,946,385	1,026,957,384	1,037,846,295	1,036,857,294

The Kurchan square

Mirror Magic Square

Those of you who enjoy playing with mirrors and kaleidoscopes will be in awe of the mirror magic square:[10]

96	64	37	45
39	43	98	62
84	76	25	57
23	59	82	78

Mirror magic square

If you reverse each of the entries you obtain another magic square. In both cases the sum for the rows, columns, and diagonals is 242:

69	46	73	54
93	34	89	26
48	67	52	75
32	95	28	87

Rorrim magic square

Isn't that a real beauty?

Magic Tesseracts

I know of no subject in mathematics that has intrigued both children and adults as much as the idea of a fourth dimension–a spatial direction different from all the directions of our normal three-dimensional space. Philosophers and parapsychologists have meditated upon this dimension that no one can point to but may be all around us. Theologians have speculated that the afterlife, heaven, hell, angels, and our souls could reside in a fourth dimension–that God and Satan could literally be lumps of hypermatter in a four-dimensional space inches away from our ordinary three-dimensional world. Throughout time, various mystics and prophets have likened our world to a three-dimensional cage and have speculated on how great our perceptions would be if we could break from the confines of our world into higher dimensions. Yet, despite all the philosophical and spiritual implications of the fourth dimension, this extra dimension also has a very practical side. Mathematicians and physicists use the fourth dimension every day in calculations. It's part of important theories that describe the very fabric of our universe.[11]

As mentioned in chapter 2, mathematician John Robert Hendricks has constructed many four-dimensional tesseracts with magic properties,[12] and you should reread this prior section for an introduction to these kinds of figures. Just as with traditional magic squares whose rows, columns, and diagonals sum to the same number, this four-dimensional analogue has the same kinds of properties in four-space. Figure 3.5 represents the projection of the four-dimensional cube onto the two-dimensional plane of the paper. Each cubical "face" of the tesseract has six 2-D faces consisting of 3×3 magic squares. (The cubes are warped in this projection in the same way that the faces of a cube are warped when drawn on 2-D paper.) To understand the magic tesseract, look at the 26 in the upper left corner. The top forwardmost edge contains 26, 60, and 37, which sum to 123. The vertical columns, such as 26, 33, and 64, sum to 123. Each oblique line of three numbers, such as 26, 67, and 30, sums to 123. A fourth linear direction shown by 26, 58, and 39 sums to 123. Can you find other magical sums? This kind of figure was first sketched in 1949. The pattern was eventually published in Canada in 1962 and later in the United States. Creation of the figure dispelled the notion that such a pattern could not be made.

If possible, study Figure 3.5 while listening to music from the *X-Files, Outer Limits,* or *Twilight Zone* TV shows—just to get you in the mood for contemplating higher dimensions.

In October 1999 John Hendricks constructed a sixth-order magic tesseract with an inlaid magic tesseract of order 3. This became the world's first inlaid magic tesseract. You can think of this smaller magic tesseract as lodged within the larger magic tesseract like a cavity sitting within a tooth or a pearl within an oyster. However, the metaphor is not quite accurate unless we can imagine a four-dimensional cavity, tooth, pearl, and oyster. In general, a tesseract can be subdivided into smaller tesseracts just as a cube may be cut into smaller cubes or a square cut into smaller squares.

The sixth-order magic tesseract contains the consecutive numbers from 1 to 1296 with the required 872 magic sums of 3891. The inlaid (internal) magic subtesseract of order 3 has the required 116 magic sums of 1824. Recall from chapter 2 that a magic tesseract contains the numbers 1 through N^4 arranged in such a way that

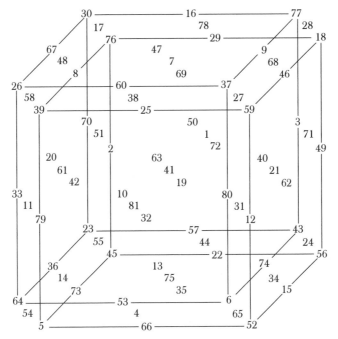

Figure 3.5
John Hendricks's magic tesseract.

the sum of the numbers in each of the N^3 rows, N^3 columns, N^3 pillars, N^3 "files" (a term used to imply a four-spatial direction), and in the eight major "quadragonals" (that pass through the center and join opposite corners) is a constant sum $\mathcal{S} = N(1 + N^4)/2$, where N is the order of the tesseract. The following array is just one magic square "side" of the inlaid third-order magic subtesseract. As you might imagine, representing tesseracts within tesseracts is a graphical nightmare, and John and I would be interested in hearing from readers who may wish to collaborate to produce insightful and attractive representations of these intricate objects. Note that the remarkable structures of John Hendricks (which are displayed throughout this book) are made all the more remarkable by the fact that he does not own a computer or a car and is in his seventies. His computations are aided by a small programmable calculator.

610	624	590
641	580	603
573	620	631

One slice of the inlaid tesseract

Jerusalem Overdrive

In this section I divert your attention to a fun class of puzzles related to magic squares and reintroduce you to the concept of Latin squares. You can consult chapter 2 for additional background information.

You are in Jerusalem, overseeing the construction of a new multidenominational religious center that would house prayer rooms for the three major local religions: Judaism (✡), Christianity (✝), and Islam (☪). To make it more difficult for terrorists to bomb or shoot a rifle through any one religious class, and to minimize religious conflicts, the architect is to design the center as a 3×3 matrix of prayer rooms so that (when viewed from above) each row and column contains only one prayer room of a particular religious denomination. An aerial view of the religious center looks like a tic-tac-toe board in which you are not permitted to have two of the same religions in any row or column. Is this arrangement possible?

The following is an arrangement prior to your attempt to minimize conflict:

✡	✡	✡
✝	✝	✝
☪	☪	☪

A dangerous situation

For a second problem, consider that you must place the prayer rooms so that each row and column contains exactly *two* religions. Is this possible?

You can design a computer program to solve this problem by representing the three religions as red, green, and amber squares in a 3×3 checkerboard. The program uses three squares of each color. Have the computer randomly pick combinations, and display them as fast as it can, until a solution is found. The rapidly changing random checkerboard is fascinating to watch, and there are quite a lot of different possible arrangements. In fact, for a 3×3 checkerboard there are 1680 distinct patterns. If it took your computer one second to compute and display each 3×3 random pattern, how long would it take, on average, to solve the problem and display a winning solution? (There is more than one winning solution.)

If you couldn't solve the first problem, work in teams until you solve it. For the second problem, here is one way to arrange the religions so that there are only two of the same religions in each row and column:

Two in each row and column

Try these problems on a few friends. Many people have difficulty visualizing the solution.

The Jerusalem overdrive problem can be thought of as a special problem in the remarkably rich mathematical area concerned with Latin squares.[13] (See chapter 2.) Latin squares were first systematically developed by Swiss mathematician Leonhard Euler in 1779. He defined a *Latin square* as a square matrix with N^2 entries of N different elements, none of which occur twice or more within any row or column of the matrix. The integer N is called the *order* of the Latin square. Recently, the subject of Latin squares has attracted the serious attention of mathematicians due to the relevance of squares to the study of combinatorics and error-correcting codes. Here's an example of the occurrence of a Latin square when considering the equation $z = (2x + y + 1)$ modulo 3:

	0	1	2	*x*
0	1	2	0	
1	0	1	2	
2	2	0	1	

y

Latin square from $z = (2x + y + 1)$ modulo 3

To understand this table, consider the example case of $x = 2$ and $y = 2$, which yields $2x + y + 1 = 7$. 7 mod 3 is 1 because $7/3$ has a remainder of 1. The 1 entry is in the last row and column of this Latin square.

Here's an interesting example of an order-10 Latin square containing two Latin subsquares of order 4 (consisting of elements 1, 2, 3, and 4) and also one of order 5 (with elements 3, 4, 5, 6, and 7), the intersection of which is a subsquare of order 2 (with elements 3 and 4):

1	9	2	8	0	6	7	4	5	3
8	2	1	0	9	7	5	3	4	6
2	1	0	9	8	5	6	7	3	4
0	8	9	1	2	3	4	6	7	5
9	0	8	2	1	4	3	5	6	7
5	6	7	3	4	1	2	0	8	9
6	7	5	4	3	2	1	8	9	0
7	4	3	5	6	0	9	1	2	8
3	5	4	6	7	8	0	9	1	2
4	3	6	7	5	9	8	2	0	1

Order-10 Latin square with Latin subsquares

Can you create Latin squares with more internal subsquares than this? What is the world record for the number of subsquares in an $N \times N$ Latin square?

A *transversal* of a Latin square of order N is a set of N cells, one in each row, one in each column, such that no two of the cells contain the same symbol. (For more on transversals, see note 13.) Even in cases where a Latin square has no transversals, the square often has partial transversals of $N-1$ elements. Do all Latin squares have a partial transversal of $N-1$ elements if the squares do not contain a true transversal? Here is an example of a Latin square with an $N-1$ transversal (I've marked the transversal path with bold cells):

1	6	3	7	4	9	2	5	0	8
2	0	4	6	5	8	3	1	9	7
3	9	5	0	1	7	4	2	8	6
4	8	1	9	2	6	5	3	7	0
5	7	2	8	3	0	1	4	6	9
6	1	8	2	9	4	7	0	5	3
7	5	9	1	0	3	8	6	4	2
8	4	0	5	6	2	9	7	3	1
9	3	6	4	7	1	0	8	2	5
0	2	7	3	8	5	6	9	1	4

A near-transversal

Now consider an amazing Latin *cube*. You can think of it as a stack of file cards. Each card contains N rows and N columns. Each number occurs exactly once in each row, once in each column, and once in each row and column in the third dimension:

0	1	2
1	2	0
2	0	1

1	2	0
2	0	1
0	1	2

2	0	1
0	1	2
1	2	0

Top section of cube   Middle section of cube   Bottom section of cube

Can you design a four-dimensional Latin hypercube? Note that computers are much faster than humans in finding errors in Latin squares, cubes, and hypercubes. So, if you are not sure if the Latin square you've written down is correct, try to write a computer program that checks each row and column. Have your computer create 4×4 Latin squares by randomly selecting values for the squares and then checking if the result is a Latin square. How long does it take your computer to find a Latin square using this trial-and-error approach? Several minutes? Hours?

A *normalized* or *reduced* Latin square is a Latin square with the first row and column given by $\{1, 2, 3, \ldots, N\}$. The numbers of normalized Latin squares of order $N = 1, 2, 3, \ldots$ are 1, 1, 1, 4, 56, 9408, ...

π-Square

Another interesting square is the enigmatic π-square invented by T. E. Lobeck of Minneapolis.[14] He starts with a conventional 5×5 magic square and then substitutes the nth digit of π (3.14159. . .) for each number n in the square. This means that a 3 is substituted for a 1, a 1 is substituted for a 2, a 4 is substituted for a 3, and so on. Amazingly, every column sum duplicates some row sum for the π-square. For example, the top row sums to 24 as does the fourth column.

17	24	1	8	15
23	5	7	14	16
4	6	13	20	22
10	12	19	21	3
11	18	25	2	9

2	4	3	6	9
6	5	2	7	3
1	9	9	4	2
3	8	8	6	4
5	3	3	1	5

5×5 magic square   π-square

Upside-Down Magic Pair

In the upside-down magic pair here, the numbers in one square will form the other square by turning either square upside down.[15] In either case, the sum of the numbers in the first place digits $(9 + 8 + 1 + 6)$ or the sum of the numbers in the second place digits $(9 + 1 + 6 + 8)$ in each row, column, 2×2 opposite short diagonal, or main diagonal will be 24. Also, in either square, the sum of the numbers in the first place digits and the sum of the numbers in the second place digits—for the four corner squares, for the four center squares, and for any four adjacent corner squares—equals 24. The sum of the numbers in each row, column, short diagonal, main diagonal, the four corner squares, the four center squares, and any four adjacent corner squares equals 264.

99	81	16	68
18	66	91	89
61	19	88	96
86	98	69	11

89	68	96	11
91	16	88	69
18	99	61	86
66	81	19	98

Upside-down of right square   Upside-down of left square

Here's a similar magic square:[16]

18	99	86	61
66	81	98	19
91	16	69	88
89	68	11	96

69-Square

Notice how the digits 0, 1, 6, 8, and 9, when rotated 180 degrees, become 6, 8, 9, 1, and 0. This magic square is still magic when rotated 180 degrees. However, if these digits are simply turned upside down, the 6 becomes a backward 9 and the 9 a backward 6. If you turn the square upside down, then reverse the 6s and 9s so they read correctly, you end up with different numbers, but the square is still magic! Notice that the corners of any 2×2, 3×3, or 4×4 square also sum to 264, as well as many other combinations.

Mirror (or IXOHOXI) Square

This square totals 19,998 in all directions in the square as is, upside down, or as reflected in a mirror.[17] In every case, any 2×2 sub-square (e.g., one formed by 8188, 1111, 1881, and 8818), as well as the four corner cells, totals 19,998.

8811	8188	1111	1888
1118	1881	8818	8181
8888	8111	1188	1811
1181	1818	8881	8118

IXOHOXI square

This novelty magic square is known as the IXOHOXI magic square. It is pandiagonal, so the four rows, four columns, two main diagonals, six complementary diagonal pairs, and sixteen 2×2 squares all sum to 19,998.

Check this out with a mirror! All numbers in the reflection will read correctly because both the 1 and the 8 are symmetric about both the horizontal and the vertical axes. Note also that the name IXOHOXI has the same characteristics when viewed in a mirror.

Meyers Eighth-Order Magic Cube

As discussed in detail in chapter 2, a magic cube is a cube composed of N^3 numbers, where N is the order of one of the squares on the side of the cube. The numbers used in forming a *perfect magic cube* are usually 1 through N^3. They are arranged so that each row, each column, each main diagonal of the square cross sections, and each of the four great (space) diagonals (sometimes called "triagonals") containing N integers will add to the same sum.

In 1970, a sixteen-year-old high school student named Richard Meyers discovered a perfect eighth-order magic cube. Meyers obtained this cube "after three months, seven theories, and thirty-one sheets of graph paper."[18] The following are eight orthogonal slices through the cube. The left column of 8×8 squares contains cross sections 1 through 4. The right column contains cross sections 5 through 8. Think of this as one huge numerical cheese chunk which you, with your quick knife, can cut to reveal the cube's internal structure.

The Meyers cube is replete with amazing symmetries:

- Every orthogonal and diagonal line of eight numbers in the Meyers cube, including the four space diagonals, adds up to 2052.
- The cube is associative; that is, any two numbers symmetrically opposite the center add up to 513.
- The eight corner cells of the little cube total 2052.
- The corners of every rectangular solid centered in the cube add up to 2052. For example, take the smallest such solid composed of the four central cells in cross section 4 (202, 123,

19	497	255	285	432	78	324	162
303	205	451	33	148	370	128	414
336	174	420	66	243	273	31	509
116	402	160	382	463	45	291	193
486	8	266	236	89	443	181	343
218	316	54	472	357	135	393	107
185	347	85	439	262	232	490	12
389	103	361	139	58	476	214	312
134	360	106	396	313	219	469	55
442	92	342	184	5	487	233	267
473	59	30	215	102	392	138	364
229	263	9	491	346	188	438	88
371	145	415	125	208	302	36	450
79	429	163	321	500	18	288	254
48	462	196	290	403	113	383	157
276	242	512	30	175	333	67	417
306	212	478	64	141	367	97	387
14	496	226	250	433	83	349	191
109	399	129	355	466	52	318	224
337	179	445	95	238	272	2	484
199	293	43	457	380	154	408	118
507	25	279	245	72	422	172	330
412	122	376	150	39	453	203	297
168	326	76	426	283	249	503	21
423	69	331	169	28	506	248	278
155	377	119	405	296	198	460	42
252	282	24	502	327	165	427	73
456	38	300	202	123	409	161	373
82	436	190	352	493	15	257	227
366	144	386	100	209	307	61	479
269	239	481	3	178	340	94	448
49	467	221	319	398	112	354	132

381	159	401	115	194	292	46	464
65	419	173	335	510	32	274	244
34	452	208	304	413	127	369	147
286	256	498	20	161	323	77	431
140	362	104	390	311	213	475	57
440	86	348	186	11	489	231	261
471	53	315	217	108	394	136	358
235	265	7	485	344	182	444	90
492	10	264	230	87	437	187	345
216	310	60	474	363	137	391	101
183	341	91	441	268	234	488	6
395	105	359	133	56	470	220	314
29	411	241	275	418	68	334	176
289	195	461	47	158	384	114	404
322	164	430	80	253	287	17	499
128	416	146	372	449	35	301	207
96	446	180	338	483	1	271	237
356	130	400	110	223	317	51	465
259	225	495	13	192	350	84	434
63	477	211	305	388	98	368	142
425	75	325	167	22	504	250	284
149	375	12	411	298	204	454	40
248	280	26	508	329	171	421	71
458	44	294	200	117	407	153	379
201	299	37	455	374	152	410	124
501	23	281	251	74	428	166	328
406	120	378	156	41	459	197	295
170	332	70	424	277	247	505	27
320	222	468	50	131	353	111	397
4	482	240	270	447	93	339	177
99	385	143	365	480	62	308	210
351	189	435	81	228	258	16	494

Slices through the Meyers cube

352, and 493) and the four central cells in cross section 5 (20, 161, 390, and 311) highlighted in the table. These cells form a $2 \times 2 \times 2$ cube at the heart of the $8 \times 8 \times 8$ cube. The sum of the corners of the little cube is 2052.

- The cube can be cut into sixty-four order-2 cubes. The eight numbers in each of these cubes add up to 2052. For example, form a cube from the two 2×2 squares, one in the upper left corner of slice 1 (19, 497, 303, and 205) and the other in the upper left corner of slice 2 (134, 360, 442, and 92). These cells are highlighted in the diagram. The sum of the eight numbers is 2052.

Think of this magic cube as a hunk of precious cheese whose internal symmetries are produced by a mathematician dairy farmer high on LSD. What other symmetries can you find?

The beautiful symmetries make possible a tantalizing number of rearrangements of the cube, all in a sense isomorphic, and every arrangement can be rotated and reflected forty-eight ways. Martin Gardner[19] asks us to imagine this cube with each of its $8 \times 8 \times 8 = 512$ cells replaced by the same cube in any of its thousands of rearrangements or orientations. In other words, you are cramming an entire cube into each cell to produce a huge new composite cube. For example, in cell 1 we put a cube that starts with 1, in cell 2 we put a cube that starts with $8^3 + 1 = 513$, in cell 3 we put a cube that starts with $(2 \times 8^3) + 1 = 1025$, and so on for the other cells. The result is a perfect magic cube of order 64, too large and time-consuming for me to type and include in this book, but hopefully you can use your imagination. Order-64 cubes in turn will build perfect magic cubes of order 512, and the same is true for all higher orders that are powers of 8.

There are several other reported discoveries of perfect eighth-order magic cubes. The first report of such a cube was published in 1875 by an unknown person in a Cincinnati newspaper.[20] Who could this brilliant, farseeing person have been? In the late 1930s,

J. Barkley Rosser and Robert J. Walker also discussed a perfect eighth-order magic cube. They also showed that perfect pandiagonal cubes exist for all orders that are multiples of 8 and all odd orders higher than 8.[21]

Perfect associative magic cubes of order 7 exist, as do perfect associative pandiagonal magic cubes of orders 9 and 11.[22] The first perfect pandiagonal cube of order 8 was constructed by C. Planck in 1905.[23] Planck considered mind-boggling "cubes" in higher dimensions. In fact, he showed that in n-space, where n is 2 or higher, the smallest perfect pandiagonal "cube" is of order 2^n, and the smallest that is also associative (symmetrically opposite pairs of numbers have the same sum) is of order $2^n + 1$. In three dimensions, the smallest perfect pandiagonal cube is of order 8, and the smallest that is also associative is of order 9. The Meyers cube is associative but not pandiagonal.

We can stretch our mind to the limit by considering four-dimensional perfect pandiagonal cubes. The smallest such cube is of order 16. The smallest that is also associative is of order 17.

Hendricks's Order-8 Inlaid Magic Cube

Figure 3.6 shows the "exterior" of an order-8 magic cube that contains an inlaid order-4 magic cube.[24] A few of the numbers are labeled for orientation purposes, and the entire cube and inlaid cube is in the following table. Each of the six windows ("cutouts") shown in Figure 3.6 holds two order-4 pandiagonal magic squares. The inner order-4 cube is *pantriagonal*, meaning that all broken triagonal pairs sum correctly to 1026. The order-8 cube uses the numbers from 1 to 8^3 and has the magic sum of 2052. Note that it is not a requirement that planar diagonals sum correctly for a cube to be considered magic, although it is possible for an order-8 cube (the smallest order perfect cube) to have this feature. John Hendricks reasons that there are 2,717,908,992 variations of this one cube, obtainable by rotations, reflections, and transformations of the components.

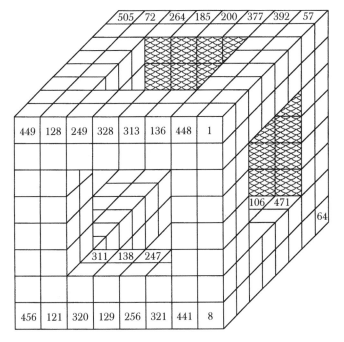

Figure 3.6
Order-8 magic cube. Contained within this cube is a hidden order-4 magic cube that has been dissected from this diagram.

On the next page are eight orthogonal slices through the cube. The left column of 8×8 squares contains cross sections 1 through 4. The right column contains cross sections 5 through 8. Note that in most cases the cross sections are only semimagic (the planar diagonals will not sum to the magic sum). The first two and last two layers of the cube contain an inlaid order-4 pandiagonal magic square.

F. Poyo's Eleventh-Order Magic Cube

In the late 1990s, Dr. F. Poyo from the Human Information Systems Laboratories at the Kanazawa Institute of Technology in

505	72	264	185	200	377	392	57
16	433	241	336	305	144	113	464
489	88	217	359	170	280	425	24
32	417	304	146	351	225	96	481
465	112	343	233	296	154	401	48
40	409	162	288	209	367	104	473
56	393	272	177	208	369	73	504
449	128	249	328	313	136	448	1
2	447	255	322	319	130	17	450
503	74	266	183	202	375	394	55
18	431	224	354	175	273	87	490
487	90	297	151	346	232	418	31
42	407	338	240	289	159	111	466
479	98	167	281	216	362	410	39
463	114	242	335	306	143	434	15
58	391	263	186	199	378	71	506
5	444	309	132	379	206	69	508
500	77	269	188	323	246	396	53
28	476	148	291	221	366	101	421
422	102	230	341	171	284	475	27
107	427	347	236	278	165	22	470
469	21	301	158	356	211	428	108
461	116	252	333	182	259	437	127
60	389	196	373	142	315	124	453
507	70	203	382	133	308	443	6
14	435	243	326	189	268	118	459
493	45	286	173	339	228	404	84
83	403	364	219	293	150	46	494
414	94	213	358	156	299	483	35
36	484	163	276	238	349	93	413
51	398	262	179	332	253	75	502
454	123	318	139	372	197	390	59

4	445	134	307	204	381	68	509
501	76	190	267	244	325	397	52
406	86	235	348	166	277	491	43
44	492	157	302	212	355	85	405
485	37	292	147	365	222	412	92
91	411	342	229	283	172	38	486
460	117	331	254	261	180	436	13
61	388	371	198	317	140	125	452
510	67	380	205	310	131	446	3
11	483	324	245	270	187	115	462
99	419	357	214	300	155	30	478
477	29	275	164	350	237	420	100
20	468	174	285	227	340	109	429
430	110	220	363	149	294	467	19
54	395	181	260	251	334	78	499
451	126	141	316	195	374	387	62
7	442	314	135	350	327	122	455
498	79	207	370	271	178	399	50
47	402	290	160	337	239	106	471
474	103	215	361	168	282	415	34
23	426	176	274	223	353	82	495
482	95	345	231	298	152	423	26
458	119	311	138	247	330	439	10
63	386	194	383	258	191	66	511
512	65	193	384	257	192	385	64
9	440	312	137	248	329	120	457
472	105	295	153	344	234	408	41
33	416	210	368	161	287	97	480
496	81	169	279	218	360	432	17
25	424	352	226	303	145	89	488
49	400	201	376	265	184	80	497
456	121	320	129	256	321	441	8

Cube within a cube

Japan studied various new cubes and methods for construction.[25]

The following are cross sections of an eleventh-order perfect magic cube from F. Poyo. The rows, columns, diagonals, and pandiagonals sum to the same value, 7326. Recall that a pandiagonal is a cyclic diagonal such as shown here:

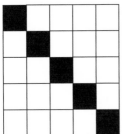  

Diagonal   Example pandiagonal   Example pandiagonal

1	1199	934	669	404	139	1326	1061	796	531	266
871	606	473	208	1274	1009	744	600	335	70	1136
410	145	1211	1078	813	548	283	18	1205	940	675
1280	1015	750	485	352	87	1153	888	623	479	214
819	554	289	24	1090	957	692	427	162	1228	1084
358	93	1159	894	629	364	231	1297	1032	767	502
1107	963	698	433	168	1234	969	836	571	306	41
646	381	237	1303	1038	773	508	243	110	1176	911
185	1251	986	842	577	312	47	1113	848	715	450
1055	790	525	260	116	1182	917	652	387	122	1320
594	329	64	1130	865	721	456	191	1257	992	727

Cross sections of Poyo's cube

134	1321	1067	802	537	272	7	1194	929	664	399
1004	739	595	341	76	1142	877	612	468	203	1269
543	278	13	1200	946	681	416	151	1217	1073	808
82	1148	883	618	474	220	1286	1021	756	491	347
952	687	422	157	1223	1079	825	560	295	30	1096
370	226	1292	1027	762	497	353	99	1165	900	635
1240	975	831	566	301	36	1102	958	704	439	174
779	514	249	105	1171	906	641	376	232	1309	1044
318	53	1119	854	710	445	180	1246	981	837	583
1188	923	658	393	128	1315	1050	785	520	255	111
716	462	197	1263	998	733	589	324	59	1125	860

267	2	1189	935	670	405	140	1327	1062	797	532
1137	872	607	463	209	1275	1010	745	601	336	71
676	411	146	1212	1068	814	549	284	19	1206	941
215	1281	1016	751	486	342	88	1154	889	624	480
1085	820	555	290	25	1091	947	693	428	163	1229
503	359	94	1160	895	630	365	221	1298	1033	768
42	1108	964	699	434	169	1235	970	826	572	307
912	647	382	238	1304	1039	774	509	244	100	1177
451	186	1252	987	843	578	313	48	1114	849	705
1310	1056	791	526	261	117	1183	918	653	388	123
728	584	330	65	1131	866	722	457	192	1258	993

Cross sections of Poyo's cube (*cont.*)

400	135	1322	1057	803	538	273	8	1195	930	665
1270	1005	740	596	331	77	1143	878	613	469	204
809	544	279	14	1201	936	682	417	152	1218	1074
348	83	1149	884	619	475	210	1287	1022	757	492
1097	953	688	423	158	1224	1080	815	561	296	31
636	371	227	1293	1028	763	498	354	89	1166	901
175	1241	976	832	567	302	37	1103	959	694	440
1045	780	515	250	106	1172	907	642	377	233	1299
573	319	54	1120	855	711	446	181	1247	982	838
112	1178	924	659	394	129	1316	1051	786	521	256
861	717	452	198	1264	999	734	590	325	60	1126

533	268	3	1190	925	671	406	141	1328	1063	798
72	1138	873	608	464	199	1276	1011	746	602	337
942	677	412	147	1213	1069	804	550	285	20	1207
481	216	1282	1017	752	487	343	78	1155	890	625
1230	1086	821	556	291	26	1092	948	683	429	164
769	504	360	95	1161	896	631	366	222	1288	1034
308	43	1109	965	700	435	170	1236	971	827	562
1167	913	648	383	239	1305	1040	775	510	245	101
706	441	187	1253	988	844	579	314	49	1115	850
124	1311	1046	792	527	262	118	1184	919	654	389
994	729	585	320	66	1132	867	723	458	193	1259

Cross sections of Poyo's cube (*cont.*)

666	401	136	1323	1058	793	539	274	9	1196	931
205	1271	1006	741	597	332	67	1144	879	614	470
1075	810	545	280	15	1202	937	672	418	153	1219
493	349	84	1150	885	620	476	211	1277	1023	758
32	1098	954	689	424	159	1225	1081	816	551	297
902	637	372	228	1294	1029	764	499	355	90	1156
430	176	1242	977	833	568	303	38	1104	960	695
1300	1035	781	516	251	107	1173	908	643	378	234
839	574	309	55	1121	856	712	447	182	1248	983
257	113	1179	914	660	395	130	1317	1052	787	522
1127	862	718	453	188	1265	1000	735	591	326	61

799	534	269	4	1191	926	661	407	142	1329	1064
338	73	1139	874	609	465	200	1266	1012	747	603
1208	943	678	413	148	1214	1070	805	540	286	21
626	482	217	1283	1018	753	488	344	79	1145	891
165	1231	1087	822	557	292	27	1093	949	684	419
1024	770	505	361	96	1162	897	632	367	223	1289
563	298	44	1110	966	701	436	171	1237	972	828
102	1168	903	649	384	240	1306	1041	776	511	246
851	707	442	177	1254	989	845	580	315	50	1116
390	125	1312	1047	782	528	263	119	1185	920	655
1260	995	730	586	321	56	1133	868	724	459	194

Cross sections of Poyo's cube (*cont.*)

932	667	402	137	1324	1059	794	529	275	10	1197
471	206	1272	1007	742	598	333	68	1134	880	615
1220	1076	811	546	281	16	1203	938	673	408	154
759	494	350	85	1151	886	621	477	212	1278	1013
287	33	1099	955	690	425	160	1226	1082	817	552
1157	892	638	373	229	1295	1030	765	500	356	91
696	431	166	1243	978	834	569	304	39	1105	961
235	1301	1036	771	517	252	108	1174	909	644	379
984	840	575	310	45	1122	857	713	448	183	1249
523	258	114	1180	915	650	396	131	1318	1053	788
62	1128	863	719	454	189	1255	1001	736	592	327

1065	800	535	270	5	1192	927	662	397	143	1330
604	339	74	1140	875	610	466	201	1267	1002	748
22	1209	944	679	414	149	1215	1071	806	541	276
881	627	483	218	1284	1019	754	489	345	80	1146
420	155	1232	1088	823	558	293	28	1094	950	685
1290	1025	760	506	362	97	1163	898	633	368	224
829	564	299	34	1111	967	702	437	172	1238	973
247	103	1169	904	639	385	241	1307	1042	777	512
1117	852	708	443	178	1244	990	846	581	316	51
656	391	126	1313	1048	783	518	264	120	1186	921
195	1261	996	731	587	322	57	1123	869	725	460

Cross sections of Poyo's cube (*cont.*)

1198	933	668	403	138	1325	1060	795	530	265	11
616	472	207	1273	1008	743	599	334	69	1135	870
144	1221	1077	812	547	282	17	1204	939	674	409
1014	749	495	351	86	1152	887	622	478	213	1279
553	288	23	1100	956	691	426	161	1227	1083	818
92	1158	893	628	374	230	1296	1031	766	501	357
962	697	432	167	1233	979	835	570	305	40	1106
380	236	1302	1037	772	507	253	109	1175	910	645
1250	985	841	576	311	46	1112	858	714	449	184
789	524	259	115	1181	916	651	386	132	1319	1054
328	63	1129	864	720	455	190	1256	991	737	593

1331	1066	801	536	271	6	1193	928	663	398	133
738	605	340	75	1141	876	611	467	202	1268	1003
277	12	1210	945	680	415	150	1216	1072	807	542
1147	882	617	484	219	1285	1020	755	490	346	81
686	421	156	1222	1089	824	559	294	29	1095	951
225	1291	1026	761	496	363	98	1164	899	634	369
974	830	565	300	35	1101	968	703	438	173	1239
513	248	104	1170	905	640	375	242	1308	1043	778
52	1118	853	709	444	179	1245	980	847	582	317
922	657	392	127	1314	1049	784	519	254	121	1187
461	196	1262	997	732	588	323	58	1124	859	726

Cross sections of Poyo's cube (*cont.*)

Poyo used his own algorithm to create this large cube. Start with a 1 at the top, as shown in the first cross section.

1. If possible, move from site to site using knight's-move jumps. This can be represented by a unit vector specifying a movement $(1, 2, 0)$ in the x, y, z direction.
2. If step 1 is not possible, move to a location by a rule in which the unit vector of the movement is $(1, 2, 2)$.
3. If the movements in steps 1 and 2 are both not possible, use the unit vector $(1, 0, 0)$ to specify the movement.

This approach seems to be guaranteed to work only for cubes that have an order that is both prime and greater than 6.

It is perhaps fortuitous that the first four dates in the first column of the top layer (871, 410, 1280, and 819) are very important ones in history:

- *A.D. 871, Medieval Europe:* King Alfred the Great of England constructs a system of government and education allowing the unification of smaller Anglo-Saxon states in the ninth and tenth centuries. Alfred is responsible for the codification of English law, public interest in local government, and the reorganization of the army.
- *A.D. 410, Rome:* The Visigoths and their German allies sack Rome and continue their search for land and provisions through southern Gaul, Spain, and Africa. Once in Africa, they take control of the Mediterranean.
- *A.D. 1280, Medieval Europe:* Eyeglasses are invented and subsequently improved in the late medieval period.
- *A.D. 819, Persia:* Persian unity begins to disintegrate with the Samanid rulers in Northern Persia.

A Magic Hypercube in the Fifth Dimension

F. Poyo constructed this $3 \times 3 \times 3 \times 3 \times 3$ magic hypercube with magic sum 366.[26] The notation (x, y, z) in the tables, for example $(0, 0, 0)$, indicates that the corresponding 3×3 square contains the nine numbers at coordinates $(x, y, z, *, *)$ in the hypercube. In other words, (x, y, z) identifies the first three of the five coordinates, and then the 3×3 square itself gives information about the other two coordinates (element i in row j is at coordinate (j, i)).

It appears that all rows and columns plus all "major" diagonals are magic in magic hyperhypercubes. There are sixteen diagonals; each one passes through the points $(0, x, y, z, w)$, $(1, 1, 1, 1, 1)$, and $(2, 2 - x, 2 - y, 2 - z, 2 - w)$, where (x, y, z, w) are each 0 or 2. For example, when $(x, y, z, w) = (0, 0, 0, 0)$ we have the values 203, 122, and 41 (sum = 366), or when $(x, y, z, w) = (0, 2, 2, 2)$ we have 167, 122, and 77 (sum = 366). All of these have 122 in the middle because they pass through the same point in the very middle of the hypercube.

203	81	82
27	109	230
136	176	54

000

63	91	212
118	239	9
185	36	145

001

100	194	72
221	18	127
45	154	167

002

75	85	206
112	233	21
179	48	139

010

94	215	57
242	3	121
30	148	188

011

197	66	103
12	130	224
157	170	39

012

88	200	78
227	24	115
51	142	173

020

209	60	97
6	124	236
151	182	33

021

69	106	191
133	218	15
164	42	160

022

***

79	83	204
110	231	25
177	52	137

100

92	213	61
240	7	119
34	146	186

101

195	70	101
16	128	222
155	168	43

102

86	207	73
234	19	113
46	140	180

110

216	55	95
1	122	243
149	189	28

111

64	104	198
131	225	10
171	37	158

112

201	76	89
22	116	228
143	174	49

120

58	98	210
125	237	4
183	31	152

121

107	192	67
219	13	134
40	161	165

122

***

84	202	80
229	26	111
53	138	175

200

211	62	93
8	120	238
147	184	35

201

71	102	193
129	220	17
166	44	156

202

205	74	87
20	114	232
141	178	47

210

56	96	214
123	241	2
187	29	150

211

105	196	65
223	11	132
38	159	169

212

77	90	199
117	226	23
172	50	144

220

99	208	59
235	5	126
32	153	181

221

190	69	108
14	135	217
162	163	41

222

A five-dimensional cube (or hyperhypercube) has thirty-two corners and eighty "faces," as you saw in the chart in chapter 2. (Please reread the tesseract information in chapter 2 to reduce confusion.) This may be difficult to see in the flat representation of numerous 3×3 squares. For example, it may appear that there are more than thirty-two corners; however, this is not the case because not all of the corners of the individual 3×3s are corners of the $3 \times 3 \times 3 \times 3 \times 3$ object. The thirty-two corners are the points of the form (a, b, c, d, e) where $a, b, c, d,$ and e are equal to 0 or 2. You can find the corners by studying the assembly of magic squares as well as the eighty faces. (A face in a $3 \times 3 \times 3 \times 3 \times 3$ object is a set of points where three of the coordinates are a fixed selection from $(0, 2)$ and the other two coordinates take on all values (x, y) with $x,$ y in $(0, 1, 2)$.) Figure 3.7 shows one of many ways of representing a five-dimensional hypercube. Just as a tesseract can be formed by dragging a cube into the fourth dimension, a five-dimensional hypercube can be created by dragging a tesseract into the fifth

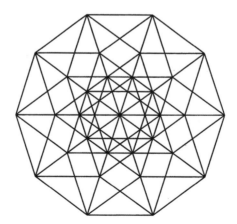

Figure 3.7
A five-dimensional hypercube.

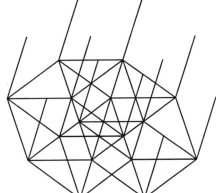

Figure 3.8
A five-dimensional hyper-cube may be created by dragging a four-dimensional tesseract into the fifth dimension.

dimension, as schematically shown in Figure 3.8. The drag lines are pointing upward and to the right.

Edwards Magic Square

The Edwards magic square,[27] named after Ronald Edwards, an amateur magician from Rochester, New York, has a magic constant of 1355. What makes the square so ingenious is that the inner 4×4 square is also a magic square in which the product of the integers in each row, column, and main diagonal is equal to 401,393,664. For example, $408 \times 122 \times 36 \times 224 = 401,393,664$. Even wilder is the fact that if all the integers in the 4×4 multiplication square are reversed (336 becomes 633 and so on) a new multiplication magic square is formed with a magic product of 4,723,906,824.

Abe Magic Square

The Abe square,[28] discovered by Gakuho Abe of Akita-ken, Japan, in 1962, is magic for both addition and multiplication. The magic constant for addition is 1200. The magic constant for multiplication is 1,619,541,385,529,760,000. To create this square, Abe used a technique given by Walter W. Horner of Pittsburgh, Pennsylvania, a retired mathematics teacher.

223	283	200	322	163	164
177	408	336	244	12	178
228	122	24	306	448	227
258	36	488	112	204	257
308	224	102	48	366	307
161	282	205	323	162	222

Edwards magic square

17	171	126	54	230	100	93	264	145
124	66	290	85	57	168	162	23	225
216	115	75	279	198	29	170	76	42
261	186	33	210	68	38	200	135	69
50	270	92	87	248	165	21	153	114
105	51	152	150	27	207	116	62	330
138	25	243	132	58	310	95	63	136
190	84	34	184	125	81	297	174	31
99	232	155	19	189	102	46	250	108

Abe magic square

Overlay Pattern Magic Square

The sixth-order overlay magic square[29] has a magic constant of
150 for the rows, columns, main diagonals, and broken diagonals.
Although it breaks the rule of a traditional magic square that would
contain the consecutive numbers 1 to 36, the square has some startling properties.

1	42	29	7	36	35
48	9	20	44	13	16
5	38	33	3	40	31
43	14	15	49	8	21
6	37	34	2	41	30
47	10	19	45	12	17

Sixth-order overlay magic square

In particular, we may overlay any of the following eight patterns on the sixth-order square so that the sum of the numbers under the dots is equal to twenty-five times the number of dots in the pattern! For example, we can place the third pattern (second row) with six dots on top of any 4×4 set of numbers in the magic square. One such overlay is highlighted in the magic square. The sum of these numbers is 150, which is twenty-five times the number of dots in the overlay pattern. Try out some of the other patterns and see if you can discover additional patterns that no one else has discovered.

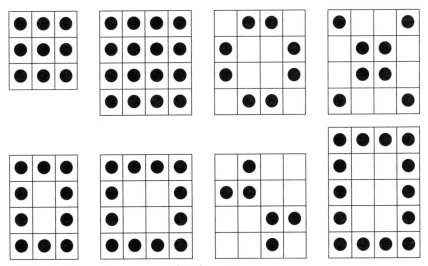

Overlay patterns

Prison Magic Squares

As with some other fantastic magic squares in the literature, the following 7×7 magic square[30] was created by a puzzle enthusiast while locked away in prison. One wonders what effect there would be on magic square research, mathematics, and society if prisoners were rewarded for any novel magic squares they created.

Correctional Populations in the United States, 1980–96: Bureau of Justice Statistics[31]

	Probation	Jail	Prison	Parole	Total
1980	1,118,097	182,288	319,598	220,438	1,840,421
1981	1,225,934	195,085	360,029	225,539	2,006,587
1982	1,357,264	207,853	402,914	224,604	2,192,635
1983	1,582,947	221,815	423,898	246,440	2,475,100
1984	1,740,948	233,018	448,264	266,992	2,689,222
1985	1,968,712	254,986	487,593	300,203	3,011,494
1986	2,114,821	272,736	526,436	326,259	3,240,252
1987	2,247,158	294,092	562,814	362,748	3,466,812
1988	2,386,427	341,893	606,810	407,596	3,742,726
1989	2,522,125	393,303	683,382	456,803	4,055,613
1990	2,670,234	403,019	743,382	531,407	4,348,042
1991	2,728,472	424,129	792,535	590,442	4,535,578
1992	2,811,611	441,781	850,566	658,601	4,762,559
1993	2,903,061	455,500	909,381	676,100	4,944,042
1994	2,981,022	479,800	990,147	690,371	5,141,340
1995	3,077,861	499,300	1,078,542	679,421	5,335,124
1996	3,180,363	510,400	1,127,528	704,709	5,523,000

(In 1999, the total number of persons in jail and prison was 6,318,900.)

As you can see from the table, in 1996 more than 5.5 million people were under some form of correctional supervision. If 3 percent of these people were willing and able to contemplate magic squares, the world would be forever changed.

The following square from prison is comprised solely of prime numbers and has a magic constant of 26,627 along every row, column, main diagonal, and broken diagonal. If you are not impressed yet, consider this: If you remove the rightmost digit from every number—for example, 5387 becomes 538—then a new magic square is created with magic constant 2760 along every row, column, main diagonal, and broken diagonal.

11	3851	9257	1747	6481	881	5399
6397	827	5501	71	3779	9221	1831
3881	9281	1759	6361	911	5417	17
839	5381	101	3797	9227	1861	6421
9311	1777	6367	941	5441	29	3761
5387	131	3821	9239	1741	6451	857
1801	6379	821	5471	47	3767	9341

Prison magic square

Some Prime Records

The following "small-constant" square on the left has the smallest possible magic constant, 177, for an order-3 square composed only of prime numbers. The number 1 is usually not considered a prime number; however, if 1 is allowed as a prime, the only all-prime order-3 magic square has a constant of 111. The square on the right[32] has the smallest possible magic constant, 3117, for an order-3 square filled with primes in an arithmetic sequence. (In an arithmetic sequence each term is equal to the sum of the preceding term and a constant. Can you figure out what the constant is for this square?)

71	89	17
5	59	113
101	29	47

World-record prime square:
smallest magic constant

1669	199	1249
619	1039	1459
829	1879	409

World-record prime arithmetic
square: smallest magic constant

What other magical properties can you discover in these two squares? Can you find similar world records for higher-order squares?

Harry Nelson was the first person to produce the following 3×3 matrix containing only consecutive primes (for which he won a $100 prize offered by Martin Gardner):[33]

1480028159	1480028153	1480028201
1480028213	1480028171	1480028129
1480028141	1480028189	1480028183

Consecutive primes

Minimum Prime Pandiagonal Order-6 Magic Square

In 1991, Allan William Johnson, Jr., of Arlington, Virginia, discovered a pandiagonal magic square consisting of thirty-six consecutive primes.[34] Here is a minimum pandiagonal order-6 magic square with numbers 67 to 251. The magic sum is 930.

67	193	71	251	109	239
139	233	113	181	157	107
241	97	191	89	163	149
73	167	131	229	151	179
199	103	227	101	127	173
211	137	197	79	223	83

Minimum prime pandiagonal square

All pandiagonal order-6 magic squares composed of thirty-six odd numbers have a magic sum of the form $6(2k + 1)$. In the previous example, $k = 77$.

Smallest Constants

The following square on the left has the smallest possible magic constant, 36, for an order-3 square composed of *composite numbers* in *arithmetic sequence*. (*Composite numbers,* as opposed to prime numbers, can be expressed as the product of other numbers, e.g., $10 = 5 \times 2$. As just mentioned, in an *arithmetic sequence* each term is equal to the sum of the preceding term and a constant.) The square[35] on the right has the smallest possible magic constant, 354, for an order-3 square composed of composite numbers in consecutive order. Can you find similar world records for higher-order squares?

18	4	14
8	12	16
10	20	6

World-record composite
arithmetic sequence:
smallest magic constant

121	114	119
116	118	120
117	122	115

World-record composite
consecutive square: smallest
magic constant

Here are fourth-order and fifth-order magic squares, discovered by Chas D. Shuldham,[36] having the lowest possible summation when made exclusively of consecutive composite numbers.

539	525	526	536
528	534	533	531
532	530	529	535
527	537	538	524

World-record composite
consecutive fourth-order square:
smallest magic constant

1328	1342	1351	1335	1344
1350	1334	1343	1332	1341
1347	1331	1340	1349	1333
1339	1348	1337	1346	1330
1336	1345	1329	1338	1352

World-record composite consecutive
fifth-order square: smallest magic
constant

One might ask what are the sets of consecutive numbers that can be used to produce Nth-order magic composite squares with the smallest possible magic constant? The following table gives such information for squares of order 6 to 12.[37]

Order	Numbers in Magic Square
6	15,684–15,719
7	19,610–19,758
8	31,398–31,461
9	155,922–156,002
10	370,262–370,361
11	1,357,202–1,357,322
12	2,010,734–2,010,877

Magic composite squares with smallest constant

A Ninth-Order Antimagic Square

As discussed in chapter 2, an *antimagic* square is an $N \times N$ array of numbers from 1 to N^2 such that each row, column, and main diagonal produces *different* sums, and the sums form a consecutive series of integers. On page 199 is a beautiful ninth-order antimagic square.[38] The sums range from 367 to 379.

I do not believe there is any systematic method by which antimagic squares can be constructed or ways in which small antimagic squares can be turned into larger ones. It would be wonderful if someone, someday discovered a way to convert a magic square into an antimagic one using a simple mathematical recipe.

A Potpourri of Antimagic Squares

On page 200 are some additional antimagic squares, which, as explained in the previous example, are a subset of heterosquares in

52	19	81	22	29	15	42	31	76
61	10	67	23	54	79	25	33	16
57	9	71	24	38	1	51	47	75
26	78	7	69	66	77	13	27	12
39	21	74	20	37	17	49	55	64
8	65	4	62	50	34	73	41	40
56	68	2	63	14	72	35	44	6
53	30	60	32	36	3	46	43	58
11	70	5	59	48	80	28	45	18

Order-9 antimagic square

which the row, column, and diagonal sums are consecutive integers. These six antimagic squares use the consecutive digits from 1 to N^2. The different sums are indicated by the rows and columns.[39]

Antimagic squares are much more difficult to construct than heterosquares. There are no order-3 antimagic squares. As the order increases, construction becomes easier, although, as I've said previously, there doesn't seem to be any standard routine available for constructing them.

Sixth-Order Talisman

Talisman squares were invented by Sidney Kravitz, a mathematician from Dover, New Jersey.[40]

A talisman square is an $N \times N$ array of numbers from 1 to N^2 in which the difference between any one number and its neighbor is greater than some given integer constant Φ. A neighboring number is defined as being in a cell horizontally, vertically, or diagonally adjacent to the current cell.

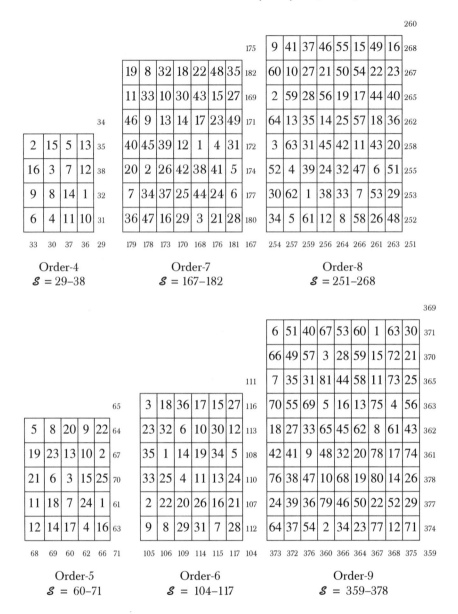

260

175

19	8	32	18	22	48	35	182
11	33	10	30	43	15	27	169
46	9	13	14	17	23	49	171
40	45	39	12	1	4	31	172
20	2	26	42	38	41	5	174
7	34	37	25	44	24	6	177
36	47	16	29	3	21	28	180

179 178 173 170 168 176 181 167

9	41	37	46	55	15	49	16	268
60	10	27	21	50	54	22	23	267
2	59	28	56	19	17	44	40	265
64	13	35	14	25	57	18	36	262
3	63	31	45	42	11	43	20	258
52	4	39	24	32	47	6	51	255
30	62	1	38	33	7	53	29	253
34	5	61	12	8	58	26	48	252

254 257 259 256 264 266 261 263 251

34

2	15	5	13	35
16	3	7	12	38
9	8	14	1	32
6	4	11	10	31

33 30 37 36 29

Order-4
$\mathcal{S} = 29{-}38$

Order-7
$\mathcal{S} = 167{-}182$

Order-8
$\mathcal{S} = 251{-}268$

369

6	51	40	67	53	60	1	63	30	371
66	49	57	3	28	59	15	72	21	370
7	35	31	81	44	58	11	73	25	365
70	55	69	5	16	13	75	4	56	363
18	27	33	65	45	62	8	61	43	362
42	41	9	48	32	20	78	17	74	361
76	38	47	10	68	19	80	14	26	378
24	39	36	79	46	50	22	52	29	377
64	37	54	2	34	23	77	12	71	374

373 372 376 360 366 364 367 368 375 359

111

3	18	36	17	15	27	116
23	32	6	10	30	12	113
35	1	14	19	34	5	108
33	25	4	11	13	24	110
2	22	20	26	16	21	107
9	8	29	31	7	28	112

105 106 109 114 115 117 104

65

5	8	20	9	22	64
19	23	13	10	2	67
21	6	3	15	25	70
11	18	7	24	1	61
12	14	17	4	16	63

68 69 60 62 66 71

Order-5
$\mathcal{S} = 60{-}71$

Order-6
$\mathcal{S} = 104{-}117$

Order-9
$\mathcal{S} = 359{-}378$

On page 201 is a sixth-order talisman. Can you determine the value of Φ used here? Can you create another sixth-order talisman with a smaller Φ?

28	10	31	13	34	16
19	1	22	4	25	7
29	11	32	14	35	17
20	2	23	5	26	8
30	12	33	15	36	18
21	3	24	6	27	9

Sixth-order talisman

Is there a practical use of talismans? Joseph Madachy, author of *Madachy's Mathematical Recreations*,[41] suggests that agricultural scientists might find some use for them when separation of mutually interfering species of plants is desired on a restricted area of highly variable soil.

Twelfth-Order Composite Square

A composite square is a magic square composed of a number of smaller magic squares. On page 202 is a big, beautiful one.[42] The entire square has magic constant 870. Each 4×4 subsquare is a perfect magic square, each with a different magic constant. Can you see how such a square might be constructed?

Order-8 Associated Square Alpha with Geometrical Diagram

The associative 8×8 magic square[43] on page 202 with magic constant 260 has a beautiful geometrical diagram (Fig. 3.9). (Notice that skew-related cells add up to the same sum, for example $29 + 36 = 8 + 57 = 33 + 32 = 65$.) The geometrical diagram makes it clear that there are three different underlying symmetries that characterize the order-8 square. Contrast this with the order-8 square in the Introduction that has two underlying symmetries.

113	127	126	116	1	15	14	4	81	95	94	84
124	118	119	121	12	6	7	9	92	86	87	89
120	122	123	117	8	10	11	5	88	90	91	85
125	115	114	128	13	3	2	16	93	83	82	96
33	47	46	36	65	79	78	68	97	111	110	100
44	38	39	41	76	70	71	73	108	102	103	105
40	42	43	37	72	74	75	69	104	106	107	101
45	35	34	48	77	67	66	80	109	99	98	112
49	63	62	52	129	143	142	132	17	31	30	20
60	54	55	57	140	134	135	137	28	22	23	25
56	58	59	53	136	138	139	133	24	26	27	21
61	51	50	64	141	131	130	144	29	19	18	32

Twelfth-order composite square

1	7	62	61	60	59	2	8
49	10	14	53	52	11	15	56
48	42	19	20	21	22	47	41
40	39	27	28	29	30	34	33
32	31	35	36	37	38	26	25
24	18	43	44	45	46	23	17
9	50	54	13	12	51	55	16
57	63	6	5	4	3	58	64

8 × 8 magic square "Alpha"

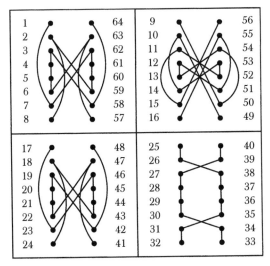

Figure 3.9
Geometric diagram for Alpha, an eighth-order magic square. Note there are three underlying symmetries made evident in this representation.

Order-8 Associated Square Beta with Geometrical Diagram

The associative 8×8 magic square[44] on page 204 with magic constant 260 has a geometrical diagram (Fig. 3.10) similar to the previous square's, except that there are four separate underlying structures rather than three. (Notice that skew-related cells add up to the same sum, for example, $29 + 36 = 8 + 57 = 33 + 32 = 65$.) The linear geometrical diagram makes it clear that there are four different underlying symmetries that characterize the order-8 square. Contrast this with the order-8 square in the Introduction with two underlying symmetries.

Hidden Symmetries via *o, ro, i,* and *ri*

It is possible to reveal hidden symmetries in magic figures by assigning numbers to four categories with a nomenclature suggested by

1	63	59	4	5	62	58	8
56	10	54	13	12	51	15	49
24	47	19	45	44	22	42	17
25	34	38	28	29	35	39	32
33	26	30	36	37	27	31	40
48	23	43	21	20	46	18	41
16	50	14	53	52	11	55	9
57	7	3	60	61	6	2	64

8 × 8 magic square "Beta"

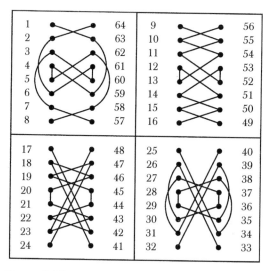

Figure 3.10
*Geometric diagram for Beta, an eighth-order magic
square.*

philosopher and mathematician Paul Carus (1852–1919),[45] who,
incidentally, wrote wonderful books on Buddhism and Buddhist
art.[46] Here is an explanation of the four categories and accompany-
ing symbols:

1. *o*–"original order," consecutive integers proceed from left to right, starting from the upper left corner. The following tables should make this clear. Numbers in this category may be represented by the symbol ■.

2. *ro*–"reversed original," consecutive integers proceed from right to left, starting from the lower right corner. In the diagrams that follow, I leave an empty cell where these occur.

3. *i*–"mirror order," consecutive integers proceed from right to left, starting from the upper right corner. This follows the directions of Hebrew, Arabic, and Farsi writing. As you can see in the diagrams, *ro* and *i* are transformed into one another by placing a vertical mirror between them. Numbers in this category may be represented by the symbol ✳.

4. *ri*–"reversed mirror," consecutive integers proceed from left to right, starting from the lower left. *ri* and *ro* are transformed into one another by placing a vertical mirror between them. (I wonder if any written languages have ever used this directional movement.) Numbers in this category may be represented by the symbol ⌘.

1	2	3	4
5	6	7	8
9	10	11	12
13	14	15	16

Order *o* (■), original

4	3	2	1
8	7	6	5
12	11	10	9
16	15	14	13

Order *i* (✳), mirror

13	14	15	16
9	10	11	12
5	6	7	8
1	2	3	4

Order *ri* (⌘),
reverse mirror

16	15	14	13
12	11	10	9
8	7	6	5
4	3	2	1

Order *ro* ( ),
reverse original

Even-order magic squares can be created by swapping cells of the *ro*, *i*, and *ri* order with the original numbers of the *o* order. It turns out that there is a beautiful symmetry that dominates these changes—a symmetry made obvious to the eye by replacing numbers with graphical symbols. You can create fourth-order squares and their multiples (e.g., 8×8, 12×12, . . .) using just *o* and *ro* cells in the array. Here is an example of a fourth-order magic square and its "Carus diagram."

1	15	14	4
12	6	7	9
8	10	11	5
13	3	2	16

Fourth-order
magic square

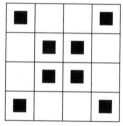

Carus diagram
showing *o* (■)
and *ro* ( )

The following is an order-12 magic square in which only *o* and *ro* numbers are used.

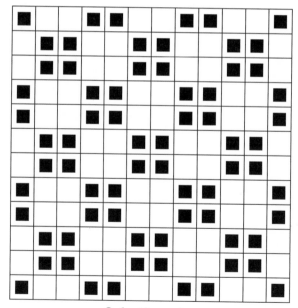

Order 12, *o* and *ro*

It's also possible to create magic squares using additional categories of numbers. For example, here are two order-8 magic squares in which all four number categories are used.

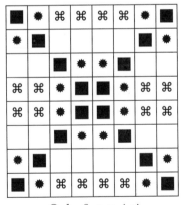

Order-8, *o, ro, i, ri*   Order-8, *o, ro, i, ri*

Here are order-8 and order-10 magic squares in which all four categories are used.

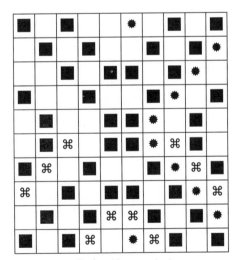

Order-8, *o, ro, i, ri*

Order-10, *o, ro, i, ri*

Paul Carus pointed out that the resulting magic patterns of *o, ro, i,* and *ri* numbers remind mathematicians of symmetrical "Chladni

Figure 3.11
Considering magic squares as acoustic waves.

patterns" produced by sand on vibrating metal plates. (Ernst Chladni was an eighteenth-century acoustician.) Carus believes that the remarkable similarities between the acoustical/mechanical patterns and the magic square structures result from analogous laws of symmetry. Carus notes, "The progressive transformations of *o*, *ro*, *i*, and *ri* by mirroring are not unlike the [sinusoidal] air waves of notes in which *o* represents the crest of the wave, *ro* the trough, *i* and *ri* the nodes."[47] Figure 3.11 illustrates the mapping of magic squares to sound waves.

A lawyer, musician, and amateur scientist, Ernst Chladni of Leipzig, Germany, found a way to make visible the vibrations caused by sound waves. He covered glass, metal, and wooden plates with sand and ran a violin bow against the plates. The vibrations moved the sand into patterns that are known today as Chladni patterns. Figure 3.12 shows Chladni's twelve engravings of these acoustically produced patterns.[48] Chladni, who was born in 1756, the same year as Mozart, and died in 1829, the same year as Beethoven, laid the foundations for acoustics, the science of sound. His experiments clearly demonstrated how sound affects matter and has the quality of creating geometric patterns.

By sprinkling sand on plates, anyone can easily create Chladni patterns in order to visualize the nodal lines of the vibrating elastic plate. The sand is thrown off the moving regions and piles up at the nodes. Normally, the plate is set vibrating by bowing it like a violin. It helps to put your fingers on the edge to select the mode you want, much like fingering the strings of a violin. You can make a modernized version of this demonstration using an electromagnetic shaker (essentially a powerful speaker) (Fig. 3.13).

Figure 3.12
Chladni patterns created by Ernst Chladni in 1787.

If we want to stretch our metaphysical musings to the extreme, there is a similarity between magic square Carus patterns and quantum particles. Both have hidden wave representations. In some sense, magic squares and quantum particles are both created and simultaneously organized by the principle of pulse or vibration. The wonder of sound, magic squares, and quantum particles is that they are not solid but rather are created by underlying waves. In an attempt to understand the dual existence of wave and physical forms, physics developed quantum field theory, in which the quantum field, or in our terminology, the vibration, is understood as the ultimate reality, and the particle or form is a manifestation of this reality. Author Cathie E. Guzzetta poetically contemplates the controversial

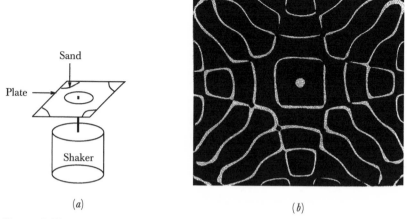

(a)

(b)

Figure 3.13

(a) Create your own Chladni patterns the modern way, with a powerful speaker (shaker).
(b) A Chladni pattern produced by the apparatus in (a). The square plate is 70 cm on a side
and made from 0.125-inch thick aluminum, painted black. The vibration is at 1450.2 Hz.
(Photo courtesy of Prof. Stephen Morris, The Department of Physics, University of Toronto.)

idea that magic squares and lifeforms are metaphors for sound: "The forms of snowflakes and faces of flowers may take on their shape because they are responding to some 'sound' in nature. Likewise, it is possible that crystals, plants, and human beings may be, in some way, music that has taken on visible form."[49]

A Very Brief but Exciting History of Knight's Tours

In chapter 2 we discussed the knight's move magic square, in which a chess knight jumps once to every square on the (8×8) chessboard in a complete tour, numbering the squares visited in order so that the magic sum is 260. In this section, I'd like to open the discussion to knight's tours in general without regard to numbering the cells upon which the knight lands. The earliest recorded solution[50] is that of Abraham De Moivre (1667–1754), the French mathematician better known for his theorems about complex numbers. Note that in De Moivre's solution (Fig. 3.14), the knight ends his tour on a square that is not one move away from the starting square. The French mathematician Adrien-Marie Legendre (1752–1833) improved on this and found a solution in which the

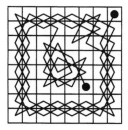

Figure 3.14
De Moivre's knight's tour.

Figure 3.15
Legendre's knight's tour.

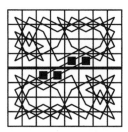

Figure 3.16
Euler's knight's tour.

first and last squares are a single move apart, so that the tour closes up on itself into a single loop of sixty-four knight's moves (Fig. 3.15). As discussed, such a tour is said to be *reentrant*. Not to be out-done, the Swiss mathematician Leonhard Euler (1707–1783) found a reentrant tour that visits two halves of the board in turn (Fig. 3.16). (The little squares show positions where the knights transit from one half to the other.)

The knight's tour can be created on boards of size 5 or greater (Fig. 3.17).[51] The tours shown on the 5 × 5 and 7 × 7 boards are not reentrant. Do you think a computer will ever find a reentrant tour on a huge 9999 × 9999 board?

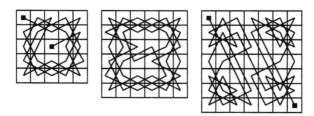

Figure 3.17
The knight's tour can be created on boards of size 5 or greater.
Shown here are order-5, 6, and 7 boards.

To answer the "9999 question," consider that a reentrant tour must visit equal numbers of black and white squares. On a 5×5 or 7×7 board (or any board with an odd number of squares altogether) a reentrant tour is therefore not possible.

A reentrant knight's tour on a 4×4 board is impossible. The great English puzzlist Henry Ernest Dudeney gives us one method to see why this is so.[52] Place a button on each square of the 4×4 board. Next, tie strings between *any* two buttons that are a knight's move apart (Fig. 3.18).

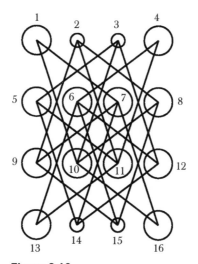

Figure 3.18
String and button model of knight's moves.

The next step is to rearrange the buttons, along with their connecting strings, so that the buttons are spread out and interconnections easier to visualize (Fig. 3.19). This spreading of the strings doesn't affect the solution, it just makes it easier to see. Let's first show that a reentrant tour is not possible. If there were such a tour, it would pass through button 13. But only buttons 6 and 11 are attached to 13, so the tour must contain either of these sequences:

6-13-11

11-13-6

Similarly, button 4 must be in the middle of either of the following sequences:

6-4-11

11-4-6

However, these sections of the supposed tour join up to form a closed loop 6-13-11-4 and back to 6. But a full tour must visit each of the 16 squares and cannot possibly contain a closed loop of length 4. Thus a reentrant quest is impossible. By careful study of

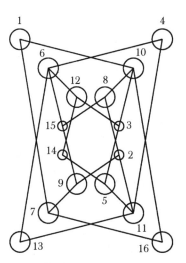

Figure 3.19
Spread string and button model of knight's moves.

the buttons and strings, you can go a step further and show that there is no possible knight's tour of a 4 × 4 square. Take a closer look at the strings and buttons with friends and see if you can understand why.

For years, people have wondered if a quest exists on the strange octagonal board shown in the following:[53]

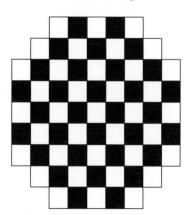

Can you find a knight's tour?

As Ian Stewart points out in *Another Fine Math You've Got Me Into,*[54] no knight's tour is possible on the diamond-shaped board. To understand why, observe that the knight changes the color of its square on each move. This means either the number of black and white squares is the same, or there is one more black than white, or one less. But the board has 32 black and 37 white squares. Thus, the longest tour that *might* be possible visits 65 squares, 32 black and 33 white, and starts and ends on a white square. (I don't know if such a tour exists.)

My favorite knight's tour is one over the six surfaces of a cube, each surface being a chessboard (Fig. 3.20a).[55] H. E. Dudeney presented this in his book *Amusement in Mathematics,* and I believe he based the solution (in which each face is toured in turn) on earlier work of French mathematician Alexandre-Théophile Vandermonde (1735–1796). Vandermonde's first love was music, and he only turned to mathematics when he was 35 years old. Vandermonde studied the theory of equations and worked on determi-

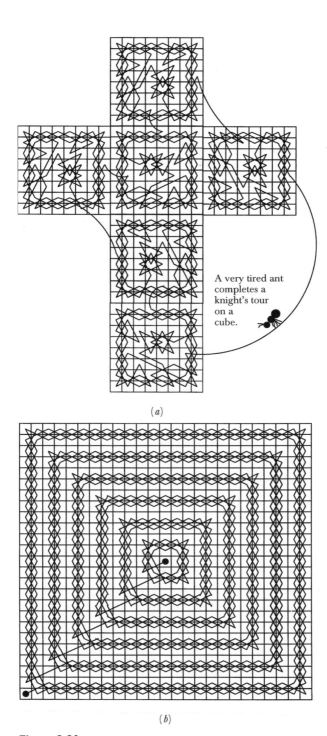

A very tired ant completes a knight's tour on a cube.

(a)

(b)

Figure 3.20

Dizzying tours. (a) Knight's moves over the surface of a cube. (b) A general method of making a tour on a board of order 4k + 1.

nants, although the determinant now named after him, by French mathematician Henri Lebesgue (1875–1941), does not appear in his published work. Vandermonde also worked on the mathematical solution of the knight's tour problem.

Another dizzying tour is shown in Figure 3.20*b*, which shows a general method of making a tour on a board of order $4k + 1$. Isn't that a beauty?

Other tours are possible on different sized cubes. On a measly $2 \times 2 \times 2$ cube, it is obviously not possible to tour each face in turn, but despite the lack of space, a reentrant tour is possible if we consider a move to be legal if, when the surface is flattened out in some way, the pattern is the same as a standard knight's move. Figure 3.21 shows a knight's move on a $2 \times 2 \times 2$ cube that may surprise you until you visualize the cube flattened out. Using these kinds of moves, you can create the reentrant tour shown in Figure 3.21*c*, in which I flattened the square for clarity. If these moves seem confusing to you, build a 3-D model and give it a try.

Knight's tours on cylinders are also possible. To visualize this we can flatten and cut the cylinder so it looks like a rectangle and place "ghost copies" of the basic rectangle at each end and pretend the corresponding cells are the same as those in the original rectangle.[56] The knight may then move off the edge onto a ghost, provided it is immediately replaced at the corresponding position of

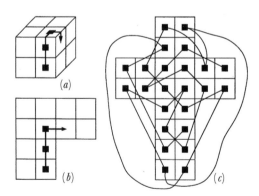

Figure 3.21
Knight's moves over a $2 \times 2 \times 2$ cube: (a) cube, (b) squashed cube showing move in (a), (c) unfolded cube showing complete tour.

the original rectangle. Tours on a $2 \times n$ cylinder or Möbius band (twisted strip) are possible only when n is odd. Tours on a $3 \times n$ and $5 \times n$ cylinder are always possible using a simple repetitive pattern. The height of such a cylinder can be any number of the form $3a + 5b$, which includes all numbers except 1, 2, 4, and 7. Even more curious is the fact that several such cylinders can be joined edge to edge and the tours combined by breaking them at suitable places and rejoining them (Fig. 3.22).

It is known that tours on a 4×4 torus exist. If a tour is possible on an $m \times n$ rectangle arranged in the form of a cylinder, it must also be possible on a torus and a Klein bottle of those dimensions. (A Klein bottle is a surface with one side, schematically illustrated in Fig. $3.22d$.)

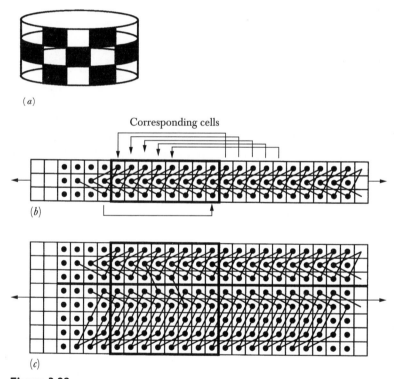

Figure 3.22
Knight's tours of cylinders. (a) A $3 \times n$ checkerboard on a cylinder. (b) Tours can be visualized by adding "ghosts" on the ends. (c) Tours on several cylinders of different sizes can be joined together by changing the links in a suitable parallelogram. (d) Schematic representation of a Klein bottle.   *(continues)*

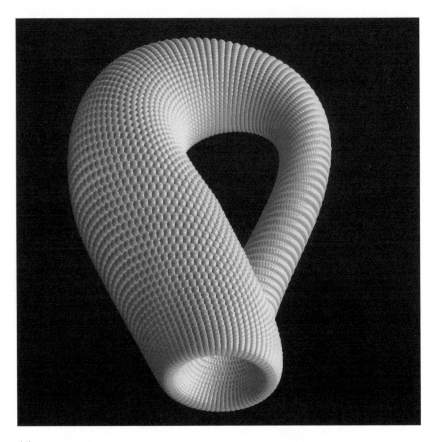

(*d*)

Figure 3.22
(continued)

"The Most Perfect Magic Square That Can Be Constructed"

In 1892, author Edward Falkener described a wonderful 8×8 magic square that he called "the most perfect magic square of 8×8 that can be constructed."[57]

Here are some of the properties of which I'm familiar:

1. The entire 8×8 array is a magic square.
2. Each quarter is an associated 4×4 square.

1	59	56	14	2	60	53	15
46	24	27	33	47	21	28	34
32	38	41	19	31	37	44	18
51	9	6	64	50	12	5	63
3	57	54	16	4	58	55	13
48	22	25	35	45	23	26	36
30	40	43	17	29	39	42	20
49	11	8	62	52	10	7	61

A supermagical eighth-order square

3. The cells in the sixteen 2×2 subsquares sum to 130.
4. Each quarter contains four 3×3 subsquares whose corner numbers sum to 130.
5. Any 5×5 square that is contained within the 8×8 square has its corner numbers in arithmetical sequence. (For example, look at the top left 5×5 square; it has corners 1, 2, 3, and 4.)

In chapter 1 we discussed how it is impossible to construct a knight's tour on a perfect magic square. Interestingly, a knight's tour can be traced through Falkener's square in such a way that every group of n numbers through the tour will have the same sum. Figure 3.23 shows such a path in which the starting numbers of each $n = 8$ period are marked. Isn't this a truly remarkable magic square?

One task that I invented a few years ago, and have not seen in the literature, is one in which you must partition the chessboard into as many as possible pieces using the paths traced out by a chess knight. In other words, think of the chessboard as a piece of cake that is subdivided into pieces by the knight's moves in Figure 3.23. What is the maximum number of pieces you can create using a knight? What is the minimum? How does your answer change for different board sizes?

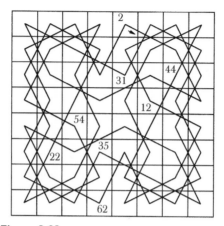

Figure 3.23
Knight's tour through the "supermagical square"
of Falkener.

The variety of patterns in the Falkener magic square motivated me to construct a geometrical diagram (Fig. 3.24). Although rather complicated looking, the geometrical diagram reveals four symmetrical "dual stars." I would be interested in hearing from any readers who can determine whether the stars have significance or yield additional insight.

Gwalior Square

In 1892, author Edward Falkener presented another wonderful 8×8 magic square that he called an Indian or Gwalior square.[58] Mr. Falkener writes

> The proper name for these squares is "Indian," for not only have the Brahmins been known to be great adepts in the formation of such squares from time immemorial, not only does [the missionary, Reverend A. H.] Frost give them an Indian name, but one of these squares is represented over the gate of Gwalior [a city in central India], while the natives of India wear them as amulets. . . .

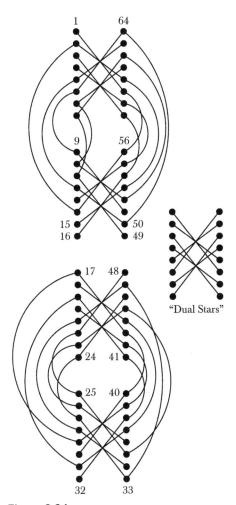

Figure 3.24
Four dual stars in Falkener's square.

In these Indian squares, it is necessary not merely that the summation of the rows, columns, and diagonals should be alike, but that *the numbers of such squares should be so harmoniously balanced that the summation of any eight numbers in one direction as in the moves of a bishop or a knight should also be alike!*

Without further ado, I introduce the Gwalior square.

1	58	3	60	8	63	6	61
16	55	14	53	9	50	11	52
17	42	19	44	24	47	22	45
32	39	30	37	25	34	27	36
57	2	59	4	64	7	62	5
56	15	54	13	49	10	51	12
41	18	43	20	48	23	46	21
40	31	38	29	33	26	35	28

Gwalior square

With a few exceptions, any eight numbers covered by a bishop's or knight's move sums to 260. Every 2×2 subsquare sums to 130. For example, $1 + 58 + 16 + 55 = 130$. These marvelous 2×2 squares can even be created by pretending the square is on the surface of a doughnut so that the ends meet. For example, $1 + 16 + 61 + 52 = 130$ and $1 + 58 + 40 + 31 = 130$. All of the square's properties remain if a row is chopped off the top and put on the bottom, or a column is chopped off the right and put on the left, and vice versa.

We've discussed this sort of square previously, but this square has additional symmetries. The first thirty-two numbers from 1 to 64 are shown on page 223. The ■ symbols mark the positions of the last thirty-two numbers. Notice that diagonally opposite quarters complement one another in the sense that entries sum to 65. Consider, for example, $1 + 64$, $58 + 7$, etc.

Frierson Square I

Frierson Square I, named after its discoverer, Mr. L. S. Frierson, was revealed to the world in the early 1900s.[59]

1	■	3	■	8	■	6	■
16	■	14	■	9	■	11	■
17	■	19	■	24	■	22	■
32	■	30	■	25	■	27	■
■	2	■	4	■	7	■	5
■	15	■	13	■	10	■	12
■	18	■	20	■	23	■	21
■	31	■	29	■	26	■	28

Numbers 1–32 for Gwalior

1	25	56	48	2	26	55	47
40	64	17	9	39	63	18	10
57	33	16	24	58	34	15	23
32	8	41	49	31	7	42	50
3	27	54	46	4	28	53	45
38	62	19	11	37	61	20	12
59	35	14	22	60	36	13	21
30	6	43	51	29	5	44	52

Frierson Square I

Here are some wonderful properties of Frierson Square I:

1. It is an order-8 magic square.
2. Each quarter is a magic square.
3. The four central rows can be split into two magic squares.
4. There are twenty-four order-3 squares, the four corner cells of which sum to 130.
5. When all the cells in *any* 4 × 4 square are summed, the sum will be 520.

6. Consider the nine centers pointed to by the arrows below:

Nine "centers"

Any rectangular parallelogram that is concentric with any of these centers contains numbers in its corner cells that sum to 130, except when the diagonals of any of the four 4 × 4 sub-squares form one side of the parallelogram. For example, 64 + 17 + 33 + 16 = 130, 40 + 9 + 57 + 24 = 130, and 40 + 25 + 41 + 24 = 130.

7. Any "unfilled octagon" of two cells on a side that is concentric with any of the nine centers has a summation constant of 260. (An example of an unfilled octagon is shown at the top of the following array.)

8. Any "filled octagon" of two cells on a side that is concentric with any of the nine centers has a summation constant of 390. (An example of a filled octagon is shown at the bottom of the following array.)

9. More than a hundred paths have been discovered that sum to 260, such as the zigzag path or bent diagonal path in the following. Other, more tortuous magic paths consist of columns of partly straight and partly zigzag movements, and partly diagonal and partly zigzag movements.

	25	56			■	■	
40			9	■			■
57			24	■			■
	8	41			■	■	
	■	■			28	53	
■	■	■	■	37	61	20	12
■	■	■	■	60	36	13	21
	■	■			5	44	

Unfilled octagon (top), filled octagon (bottom)

↘							↙
	↙					↙	
↘					↙		
	↙			↓			
↘				↘			
	↙				↘		
↘						↘	
	↘						↘

Unusual paths

Frierson Square II

Frierson Square II, also named after its discoverer, Mr. L. S. Frierson, shocked the world in 1910.[60]

P_1

64	57	4	5	56	49	12	13
3	6	63	58	11	14	55	50
61	60	1	8	53	52	9	16
2	7	62	59	10	15	54	51
48	41	20	21	40	33	28	29
19	22	47	42	27	30	39	34
45	44	17	24	37	36	25	32
18	23	46	43	26	31	38	35

P_2

Frierson Square II

The square exhibits many of the interesting properties of the previously described Franklin and Gwalior squares with respect to bent and continuous diagonals, in addition to a diverse panoply of other properties:

1. *Any* 2×2 square has a constant summation of 130, with four exceptions. (Can you find the exceptions?) For example, $64 + 57 + 3 + 6 = 130$ and, similarly, $57 + 4 + 6 + 63 = 130$. This property can be demonstrated more clearly by moving a 2×2 cutout over the square and summing the contents within.
2. The corner cells of any 3×3 square that lies entirely to the right or left of the central dividing line (designated P_1-P_2) sum to 130.
3. The corner cells of any 2×4, 2×6, or 2×8 rectangle perpendicular to the line P_1-P_2 and symmetrical therewith sum to 130. For example, $10 + 51 + 40 + 29 = 130$.
4. The corner cells of any 2×7, 3×6, or 2×8 rectangle diagonal to the line P_1-P_2 sum to 130. For example, $49 + 16 + 19 + 46 = 130$ (highlighted).
5. The corner cells of any 5×5 square contain numbers in arithmetical progression—for example, 40, 48, 56, and 64.

6. Any "constructive" diagonal sums to 260, such as the one that starts $38 + 32$, and then continues at left, $19 + 41 + 62 + 8 + 11 + 49$.

7. Any bent diagonal sums to 260 (see the Franklin square for examples of bent diagonals).

8. By subdividing the square into quarters, and subdividing each quarter into four 2×2 squares, the numbers will be symmetrically arranged in relation to cells that are similarly located in diagonally opposite 2×2 squares in each quarter. For example, $64 + 1 = 65$, $57 + 8 = 65$, and $5 + 60 = 65$.

9. Any reflected diagonal sums to 260. For example, the upper reflected bent diagonal of $64 + 6 + 1 + 58 + 11 + 52 + 55 + 13 = 260$.

Reflected bent diagonals

Frierson Square III

Frierson Square III[61] exhibits many of the features of the Gwalior and previous Frierson squares with respect to bent and continuous diagonals, but it also has an interesting constructive pattern in which the 2×2 subsquares contain consecutive numbers, and the first thirty-two numbers fall within the pleasing symmetrical pattern shown here.

1	2	44	43	21	22	64	63
3	4	42	41	23	24	62	61
56	55	52	51	13	14	9	10
54	53	50	49	15	16	11	12
25	26	5	6	60	59	40	39
27	28	7	8	58	57	38	37
48	47	29	30	36	35	17	18
46	45	31	32	34	33	19	20

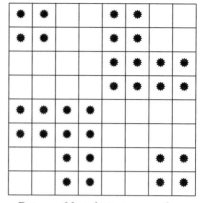

Frierson Square III   Pattern of first thirty-two numbers

Magic Cross

The Frierson Cross[62] (Fig. 3.25) exhibits many unique properties. Researchers have suggested that it contains an amazing 160,144 different "tracks" of twenty-one numbers that sum to 1471. For example, starting at the upper left we can create the sum: $2 + 91 + 128 + 73 + 3 + 90 + 129 + 72 + 1$ (central circle) $+ 62 + 139 + 80 + 13 + 63 + 138 + 81 + 12 + 64 + 137 + 82 + 11 = 1471$. Here's another example starting at the left cross and then moving downward once the central 1 is encountered: $4 + 58 + 125 + 107 + 5 + 59 + 124 + 106 + 1$ (central circle) $+ 62 + 139 + 80 + 13 + 63 + 138 + 81 + 12 + 64 + 137 + 82 + 11 = 1471$. What other magic paths can you find?

Magic Rhombus of Diocletian

The Rhombus of Diocletian[63] (Fig. 3.26) exhibits many unique properties. (A rhombus is a quadrilateral with four equal sides, usually without right-angled corners.) Here are some of the magic characteristics:

1. Each 4×4 rhombus has a magic sum of 162. For example, looking at the upper rhombus we find $61 + 60 + 40 + 1 = 162$ and $60 + 31 + 30 + 41 = 162$.

2. Each 4×4 rhombus contains four 2×2 rhombi whose contents sum to 162. For example, $61 + 60 + 10 + 31$ is a 2×2 subrhombus whose contents sum to 162. The 2×2

Figure 3.25
Frierson cross.

rhombi in the center of the 4×4 rhombi also have this characteristic. For example, $31 + 51 + 30 + 50 = 162$.

3. Each 4×4 rhombus contains four 3×3 rhombi whose corner numbers sum to 162. For example, $61 + 40 + 11 + 50 = 162$.

4. The numbers in a particular 4×4 rhombus end in only one of two possible numbers: 0 or 1 (top rhombus), 2 or 9 (right, upper rhombus), 3 or 8 (right, lower rhombus), 4 or 7 (left, lower rhombus), and 5 or 6 (left, upper rhombus).

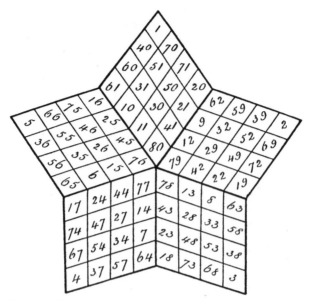

Figure 3.26
Rhombus of Diocletian.

5. The central "star" contains the consecutive numbers 76, 77, 78, 79, and 80. The outer tips of each rhombus contain the consecutive numbers 1, 2, 3, 4, and 5.

What other patterns can you find that no one else has yet discovered?

Frierson Composite Square

This ninth-order magic square by Frierson[64] is composite and contains several oddly nested internal squares. Created in the early 1900s, the 9×9 square has magic constant 369. The 7×7 square has magic constant 287. The 5×5 square has magic constant 205. The 3×3 square has magic constant 123. The differences between these successive magic constants is a constant, 82.

Interestingly, it was not until 1992 that someone noticed a minor flaw in this square. In particular, John Hendricks discovered that the seventh-order inlaid square is only semimagic. The diagonal from 49 to 80 sums to 322, but it must sum to 287 to make this square magic. It is possible to make corrections to this square to solve this problem. Can you solve the mystery and make the correction? Note 64 gives you the solution.

71	1	51	32	50	2	80	3	79
21	41	61	56	26	13	69	25	57
31	81	11	20	62	65	17	63	19
34	40	60	43	28	64	18	55	27
48	42	22	54	39	75	7	10	72
33	53	15	68	16	44	58	77	5
49	29	67	14	66	24	38	59	23
76	4	70	73	8	37	36	30	35
6	78	12	9	74	45	46	47	52

Frierson composite square

Diamond in the Rough

The internal diamond square by Mr. Frierson has beautiful properties:[65]

1. All the odd numbers are trapped within a diamond enclosure. All the even numbers are outside the enclosure.
2. Every pair of numbers located equally above and below a central horizontal axis ends in the same integer, for example 42 and 32, or 35 and 15.
3. The sum of any pair of numbers located equally right and left of the central vertical axis ends with 2.

42	58	68	64	1	8	44	34	50
2	66	54	45	11	77	78	26	10
12	6	79	53	21	69	63	46	20
52	7	35	23	31	39	67	55	60
73	65	57	49	41	33	25	17	9
22	27	15	43	51	59	47	75	30
62	36	19	13	61	29	3	76	70
72	56	4	5	71	37	28	16	80
32	48	38	74	81	18	14	24	40

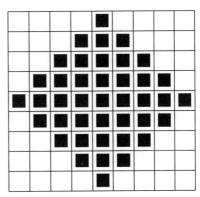

Diamond in the rough   The diamond

4. The twenty-five odd numbers in bold make a 5×5 magic square with magic sum 205. For example, $73 + 7 + 79 + 45 + 1 = 205$ and $1 + 77 + 63 + 55 + 9 = 205$.

5. The sixteen odd numbers between the bold numbers in the diamond form a 4×4 magic square.

6. The 9×9 square is associated; that is, skew cells sum to the same constant. For example, $42 + 40 = 66 + 16 = 33 + 49 = 82$.

Large, Perfect Knight's Move Magic Square

Here is a 16×16 knight's move magic square with magic constant 2056.[66] The chess knight starts at the 1 and ends its magic tour at 256. Each square is traversed only once. The tour is closed because the first and last moves are also connected by a knight move. Such a magic tour would be impossible for odd N and for an ordinary 8×8

184	217	170	75	188	219	172	77	228	37	86	21	230	39	88	25
169	74	185	218	171	76	189	220	85	20	229	38	87	24	231	40
216	183	68	167	222	187	78	173	36	227	22	83	42	237	26	89
73	168	215	186	67	174	221	190	19	84	35	238	23	90	41	232
182	213	166	69	178	223	176	79	226	33	82	31	236	43	92	27
165	72	179	214	175	66	191	224	81	18	239	34	91	30	233	44
212	181	70	163	210	177	80	161	48	225	32	95	46	235	28	93
71	164	211	180	65	162	209	192	17	96	47	240	29	94	45	234
202	13	126	61	208	15	128	49	160	241	130	97	148	243	132	103
125	60	203	14	127	64	193	16	129	112	145	242	131	102	149	244
12	201	62	123	2	207	50	113	256	159	98	143	246	147	104	133
59	124	11	204	63	114	1	194	111	144	255	146	101	134	245	150
200	9	122	55	206	3	116	51	158	253	142	99	154	247	136	105
121	58	205	10	115	54	195	4	141	110	155	254	135	100	151	248
8	199	56	119	6	197	52	117	252	157	108	139	250	153	106	137
57	120	7	198	53	118	5	196	109	140	251	156	107	138	249	152

♞ 16×16 closed magic knight's tour ♞

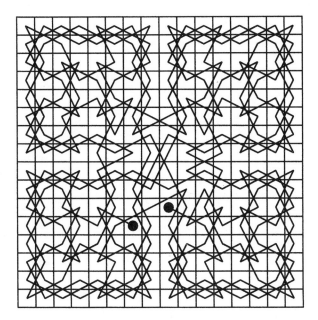

Figure 3.27
16 × 16 closed perfect magic knight's tour.

chessboard. As discussed in chapter 2, the best that can be achieved on an order-8 board is a semimagic square in which the rows and columns, but not main diagonals, have the magic constant of 260.

Figure 3.27 graphically depicts the 16 × 16 closed magic knight's tour. On page 234 are the first thirty-one moves to give you an idea how the order-16 square begins to fill with traversed cells.

King's Tour Magic Square

A magic tour for a king's moves is illustrated at the bottom of page 234 on an 8 × 8 magic square with magic sum of 260.[67] This means that it is possible to traverse the square, starting at ♔ (square 1) and ending at 64, without going through the same cell more than once. The movements used are those of a king in chess. You can trace the path just by following consecutive numbers, 1, 2, 3, 4, 5, 6, . . . , 64. Figure 3.28 graphically depicts the king's path.

184	217	170	75	188	219	172	77	228	37	86	■	230	39	88	■
169	74	185	218	171	76	189	220	85	■	229	38	87	■	231	40
216	183	68	167	222	187	78	173	36	227	■	83	42	237	■	89
73	168	215	186	67	174	221	190	■	84	35	238	■	90	41	232
182	213	166	69	178	223	176	79	226	33	82	■	236	43	92	■
165	72	179	214	175	66	191	224	81	■	239	34	91	■	233	44
212	181	70	163	210	177	80	161	48	225	■	95	46	235	■	93
71	164	211	180	65	162	209	192	■	96	47	240	■	94	45	234
202	■	126	61	208	■	128	49	160	241	130	97	148	243	132	103
125	60	203	■	127	64	193	■	129	112	145	242	131	102	149	244
■	201	62	123	■	207	50	113	256	159	98	143	246	147	104	133
59	124	■	204	63	114	■	194	111	144	255	146	101	134	245	150
200	■	122	55	206	■	116	51	158	253	142	99	154	247	136	105
121	58	205	■	115	54	195	■	141	110	155	254	135	100	151	248
■	199	56	119	■	197	52	117	252	157	108	139	250	153	106	137
57	120	■	198	53	118	■	196	109	140	251	156	107	138	249	152

The beginning moves

61	62	63	64	♚	2	3	4
60	11	58	57	8	7	54	5
12	59	10	9	56	55	6	53
13	14	15	16	49	50	51	52
20	19	18	17	48	47	46	45
21	38	23	24	41	42	27	44
37	22	39	40	25	26	43	28
36	35	34	33	32	31	30	29

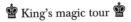 King's magic tour ♚

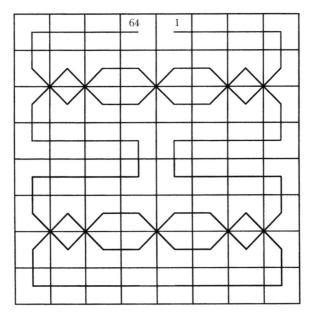

Figure 3.28
King's tour magic square.

More Tours: Visit the Bizarre Chess Leapers

Ed Pegg, Jr., is a cruciverbalist from Colorado and a former computer programmer at NORAD, where he tested his programs by running nuclear attack scenarios. He and his colleagues were stimulated by the various chess variations described in my book *Mazes for the Mind.* What happens if new pieces are introduced that move in unusual ways? For example, in chess, the knight makes an L-shaped move of one square in one direction and two squares in a perpendicular direction over squares that may be occupied. As such, the knight may be called a one-two leaper or (1, 2) leaper. What other kinds of leapers can be used in chess?

A tour of the chessboard by a knight occurs when the knight visits each square just once. The famous Swiss mathematician Leonhard Euler (1707–1783) studied these tours extensively. However, there are strange and exotic leapers to consider in addition to the traditional knight. For example, the zebra is a (2, 3) leaper. The giraffe is a $(1, 4)$ leaper. Can either of these pieces, or any other (m, n) leaper, make a tour of any board of any size? Here are some solutions:

1	26	45	68	83	16	5	24	31	14
28	47	74	43	70	3	22	53	7	33
									2
67	84	55	0	25	30	13	82	17	6
44	69	2	27	46	73	32	15	4	23
75	42	29	48	85	54	71	34	21	52
56	99	66	93	58	79	18	7	12	81
97	90	39	64	95	20	9	88	3	60
									7
92	49	76	41	62	11	86	51	7	35
									8
65	94	57	98	89	38	59	80	19	8
40	63	96	91	50	77	36	61	10	87

Closed zebra tour of 10×10 found by John Scholes

5	66	19	54	75	42	29	38	3
76	43	28	37	4	65	20	55	74
23	14	71	62	49	34	79	10	25
48	33	80	9	24	15	70	61	50
67	6	53	18	45	30	41	2	39
44	77	36	27	12	21	64	73	56
13	22	63	72	57	78	35	26	11
58	47	32	81	8	69	16	51	60
7	68	17	52	59	46	31	40	1

Open giraffe tour of 9×9 found by Juha Saukkola

8	37	58	47	30	5	36	61
45	52	21	2	13	44	51	28
64	15	24	41	54	33	16	25
39	56	49	18	9	38	57	48
12	31	4	27	62	11	22	3
7	34	59	46	29	6	35	60
42	53	20	1	14	43	50	19
63	10	23	40	55	32	17	26

Closed Pythagoron tour of 8×8 found by Juha Saukkola

49	12	51	16	41	18	45	30
40	19	44	29	48	9	62	15
59	8	63	2	39	20	35	26
38	23	32	25	58	5	56	3
11	50	13	52	17	42	31	46
*	43	28	47	10	61	14	53
7	60	1	54	21	36	27	34
22	37	24	33	6	57	4	55

Almost giraffe tour of 8×8 found by Juha Saukkola

A Pythagoron is a $(3, 4)$ leaper (antelope) plus an $(0, 5)$ leaper.

Dan Cass has proven that an 11×11 zebra tour (open or closed) is impossible. Proving that the zebra cannot tour the 8×8 or 9×9 boards is an interesting avenue of research. Ed Pegg, Jr., has proven that a 12×12 zebra tour is impossible. Juha Saukkola is almost certain that the 8×8 giraffe tour is impossible. Donald Knuth, Professor Emeritus at Stanford University, has looked at the minimum boards needed for an (m, n) leaper to travel from one cell to any other cell. Such traveling is possible only if m and n are relatively prime and of different parity. (Two integers are relatively prime if they have no prime factors in common. The numerator and denominator of a reduced fraction are relatively prime.) The smallest board is $m + n$ by $2 \times (\min(m,n))$.

Here are some unsolved problems:

1. Do tours exist for the following leapers: (1,6), (2,5), (2,7), (3,4), (3,6), (4,5), (4,7), (5,6), or (6,7)?
2. Is there an open zebra tour for the 9×10 board? (Ed Pegg, Jr., proved that the closed tour was impossible.)
3. Is there a giraffe tour on a board smaller than 9×9?
4. For a general (m,n) leaper, what is the smallest board it can tour?
5. Is there an (m,n) leaper (such that $\mathrm{GCD}(m,n) = 1$ and $m + n$ is odd) that cannot tour a board of any size? GCD stands for the greatest common divisor.

Additional information on leapers can be found at various web sites listed in note 68.

A Magic Rook's Tour

As stated several times in this book, a *tour* of a given chess piece is a sequence of moves such that each square of the chessboard is occupied once and only once during the journey. The tour is said to be *reentrant* if the chess piece can move from the last square of the tour directly back to the first square of the tour. Tours are possible by the king, queen, rook, and knight, but not by the bishop.

In the following diagram, the goal is to number the occupied chess squares with consecutive integers starting at the ♖ symbol (square 1). The tour is a magic tour if the resulting numbered array is a magic square. The tour is a semimagic tour if the result is only a semimagic square (rows and columns add up to the magic constant, but the two diagonals do not both add up to the magic constant). A previous section demonstrated a reentrant magic king's tour on the standard 8×8 chessboard. There is also a reentrant magic queen's tour.

The first magic rook's tour was discovered in 1985 by Stanley Rabinowitz of Alliant Computer Systems Corporation in Acton, Massachusetts:[69]

61	62	63	64	♜	2	3	4
12	11	10	9	56	55	54	53
20	19	18	48	17	47	46	45
60	59	58	8	57	7	6	5
37	38	39	25	40	26	27	28
13	14	15	49	16	50	51	52
21	22	23	24	41	42	43	44
36	35	34	33	32	31	30	29

♜ Magic rooks' tour ♜

The Intrinsic Harmony of Numbers

Author W. S. Andrews described the following 18×18 magic square by Harry A. Sayles with the following uplifting words:

> Considering its constructive origin and interesting features, this square, notwithstanding its simplicity, may be fairly said to present one of the most remarkable illustrations of the intrinsic harmony of numbers.[70]

To generate this square, simply write down the decimal expansion for 1/19, 2/19, 3/19, . . . , 18/19 in each row. For example, 1/19 = 0.052631578947368421. The magic square thus created has 81 as a magic constant for the rows, columns, and main diagonals. (Of course, this is not a pure magic square because a consecutive series of numbers from 1 to *N* is not used.) Look carefully at the sequence of digits in each row. Notice that these digits are the same except the position of the decimal point is changed in relation to the rest of the numbers.

The decimal representation for $n/19$ has some interesting properties. When the digits are divided by 2, many of the remaining digits are generated in order. For example

$$\frac{052631578947368421\ldots}{2} = 2631578947368421\ldots$$

$$\frac{105263157894736842\ldots}{2} = 52631578947368421\ldots$$

1/19 =	0.0	5	2	6	3	1	5	7	8	9	4	7	3	6	8	4	2	1
2/19 =	0.1	0	5	2	6	3	1	5	7	8	9	4	7	3	6	8	4	2
3/19 =	0.1	5	7	8	9	4	7	3	6	8	4	2	1	0	5	2	6	3
4/19 =	0.2	1	0	5	2	6	3	1	5	7	8	9	4	7	3	6	8	4
5/19 =	0.2	6	3	1	5	7	8	9	4	7	3	6	8	4	2	1	0	5
6/19 =	0.3	1	5	7	8	9	4	7	3	6	8	4	2	1	0	5	2	6
7/19 =	0.3	6	8	4	2	1	0	5	2	6	3	1	5	7	8	9	4	7
8/19 =	0.4	2	1	0	5	2	6	3	1	5	7	8	9	4	7	3	6	8
9/19 =	0.4	7	3	6	8	4	2	1	0	5	2	6	3	1	5	7	8	9
10/19 =	0.5	2	6	3	1	5	7	8	9	4	7	3	6	8	4	2	1	0
11/19 =	0.5	7	8	9	4	7	3	6	8	4	2	1	0	5	2	6	3	1
12/19 =	0.6	3	1	5	7	8	9	4	7	3	6	8	4	2	1	0	5	2
13/19 =	0.6	8	4	2	1	0	5	2	6	3	1	5	7	8	9	4	7	3
14/19 =	0.7	3	6	8	4	2	1	0	5	2	6	3	1	5	7	8	9	4
15/19 =	0.7	8	9	4	7	3	6	8	4	2	1	0	5	2	6	3	1	5
16/19 =	0.8	4	2	1	0	5	2	6	3	1	5	7	8	9	4	7	3	6
17/19 =	0.8	9	4	7	3	6	8	4	2	1	0	5	2	6	3	1	5	7
18/19 =	0.9	4	7	3	6	8	4	2	1	0	5	2	6	3	1	5	7	8

The intrinsic harmony of numbers

Notice that if the first nine digits of 0.052631578947368421 are added to the last nine digits, we get $052631578 + 947368421 = 81$, the magic constant. What other remarkable properties of this square can you find?

Hidden 37s

The following 6×6 magic squares possess a number of hidden symmetries in which pairs of numbers sum to 37.[71]

1	26	27	12	9	36
35	25	28	11	10	2
3	23	21	14	16	34
33	22	24	15	13	4
20	8	5	30	31	17
19	7	6	29	32	18

Magic square α

1	35	34	3	32	6
30	8	28	27	11	7
24	23	15	16	14	19
13	17	21	22	20	18
12	26	9	10	29	25
31	2	4	33	5	36

Magic square β

1	26	28	11	9	36
35	25	27	12	10	2
3	23	21	14	16	34
33	22	24	15	13	4
20	8	6	29	31	17
19	7	5	30	32	18

Magic square γ

The pair connection diagrams in Figure 3.29 indicate which number pairs have these properties. Figure 3.30 is a geometrical diagram that shows column connections and reveals additional symmetries. (Geometrical diagrams are described in the Introduction, and pair connection diagrams are described in chapter 2.)

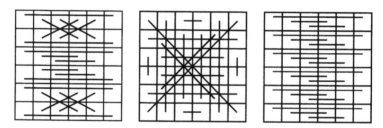

Figure 3.29
Pair connection diagram for magic squares α, β, and γ.

Wonderful Overlapping Fourteenth-Order Square with Inlays

The following 14×14 magic square has a magic constant of 1379.[72] There are two overlapping 8×8 magic subsquares in the northwest and southeast corners. There are two 6×6 magic subsquares in the northeast and southwest corners with magic constant 591. There are also many bent diagonals and zigzag rows of eight numbers that sum to 788. As just one example, start at the northwest 47

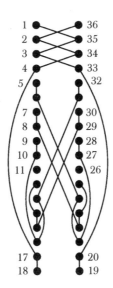

Figure 3.30
Geometrical diagram for magic square γ.

47	143	65	131	56	142	44	154	7	13	193	4	185	184
48	150	66	132	55	141	43	153	186	6	187	194	1	17
57	139	39	157	50	148	62	136	9	14	183	8	181	195
58	140	40	158	49	147	61	135	188	16	13	190	182	2
145	51	133	64	138	60	160	38	12	196	10	3	191	179
146	52	134	64	137	59	159	37	189	180	5	192	11	14
143	53	155	41	152	46	130	68	108	90	103	93	115	81
144	54	156	42	151	45	129	67	107	89	104	94	116	82
25	36	175	22	167	166	99	97	121	75	126	72	114	84
168	24	169	176	19	35	100	98	122	76	125	71	113	83
27	33	165	26	163	177	73	123	85	111	96	102	78	120
170	34	31	172	164	20	74	124	86	112	95	101	77	119
30	178	28	21	173	161	91	105	79	117	70	128	88	110
170	162	23	174	29	32	92	106	80	118	69	127	87	109

Overlapping and inlaid magic squares

and zigzag as follows: $47 + 150 + 65 + 132 + 56 + 141 + 44 + 153 = 788$. Also consider the bent diagonal: $47 + 150 + 39 + 158 + 49 + 148 + 43 + 154 = 788$.

The 6×6 square in the northeast corner has an interesting *inlaid magic square*. Like electrons collapsing into an atom's nucleus, you can remove the twenty numbers around the central 2×2 block and bring the three at each corner-angle together to form a fourth-order magic square with magic sum 394. This jigsaw relationship also holds for the southeast corner.

John R. Hendricks's Diamond-Inlay Magic Square

The following order-9 magic square has three inlaid magic squares of orders 3, 5, and 7.[73]

32	6	22	13	51	67	58	78	42
28	14	70	20	17	66	56	44	54
48	80	9	77	43	75	1	2	34
47	64	37	57	63	21	27	18	35
33	11	23	3	41	79	59	71	49
53	8	55	61	19	25	45	74	29
52	72	81	7	39	5	73	10	30
36	38	12	62	65	16	26	68	46
40	76	60	69	31	15	24	4	50

Diamond inlaid, bordered square

The order-3 magic square is rotated 45 degrees (shown in gray) and is referred to as a diamond inlay. Note that small and large numbers are mixed throughout the inner square, which is often not the case for older, traditional bordered squares, such as those in chapter 2. The magic constants are $\mathcal{S}_3 = 123$, $\mathcal{S}_5 = 205$, $\mathcal{S}_7 = 287$, and $\mathcal{S}_9 = 369$. The numbers used range from 1 to 81, so the order-9 square is a pure magic square.

Order-20 Magic Square with Inlays

The square in this section is replete with awesome, inlaid structures.[74]

Harvey Heinz assembled this order-20 masterpiece using designs from John Hendricks. Hendricks provided the overall layout and framework and four of each of the order-7 inlays, one for each quadrant. It was then a matter of deciding which type of inlay to put in each quadrant. The order-7 magic square (upper right corner) is pandiagonal so the square may be altered by shifting rows or columns. The order-5 magic square (lower left quadrant) is also pandiagonal. The order-20 square, because it contains the consecutive numbers from 1 to 400, is a pure magic square. The diamond inlay at the lower right has magic sum 570. Here are some sums for the inlaid bordered squares: upper left, 1477, 1055, 633; upper right, 1337; lower left, 1470, 1050; and lower right, 1330, 950, 570.

400	9	16	13	18	2	7	4	10	6	395	391	397	394	399	383	388	385	12	381
161	232	225	228	223	239	234	237	231	235	166	170	164	167	162	178	17	176	229	180
301	92	219	83	57	379	323	45	371	95	315	357	199	23	125	74	31	248	312	81
241	152	263	214	157	268	145	271	159	155	255	34	131	68	317	259	34	185	252	141
341	52	368	88	205	337	91	334	54	55	355	79	303	245	354	191	28	137	352	41
21	372	59	274	97	211	325	148	363	375	35	251	348	197	39	123	65	314	32	361
121	272	143	328	33	85	217	94	279	275	135	183	25	134	71	308	25	359	132	261
61	332	374	151	265	154	277	208	48	335	75	128	77	319	243	345	19	31	72	321
181	212	51	339	365	43	99	377	203	215	195	305	254	351	188	37	13	63	192	201
101	292	285	288	283	299	294	297	291	295	115	111	117	114	119	103	108	105	112	281
300	109	296	293	298	282	287	284	290	286	106	110	104	107	102	118	113	116	289	120
220	189	202	98	44	362	338	56	370	206	186	182	318	344	22	78	35	30	209	200
340	69	278	204	270	336	87	153	142	326	66	138	316	244	130	184	76	242	329	80
280	129	373	327	93	144	210	276	47	266	126	33	73	253	67	247	310	347	269	140
380	29	42	150	216	267	333	84	378	366	26	342	187	124	190	256	193	38	369	40
60	349	158	273	324	90	156	207	262	46	346	258	70	133	313	127	30	122	49	360
160	249	367	96	147	213	264	330	53	146	246	27	304	196	250	136	64	353	149	260
100	309	50	322	376	58	82	364	218	86	306	350	62	36	358	302	24	198	89	320
221	172	236	233	238	222	227	224	230	226	175	171	177	174	179	163	168	165	169	240
20	389	5	8	3	19	14	17	11	15	386	390	384	387	382	398	393	396	392	1

Order-20 magic square with inlays

In the Introduction we discussed the early magic square research done by French mathematician Bernard Frénicle de Bessy (1602–1675). Can you imagine what Frénicle would have thought of this ornate magic square produced in the 1990s? Frénicle's jaw would have dropped in amazement. His heart would have raced.

A century from now, what magic squares will make your jaded heart jump with joy?

Bimagic Square of Order 9

As we discussed in chapter 2, a bimagic square is a magic square in which you can square all the elements and the resultant square is also magic. Here is probably the first *odd-ordered* bimagic square ever discovered.[75] Its history is interesting. In 1991, David M. Collison sent this to magic square expert John Hendricks without explanation and then promptly died. No one knows for sure how Collison created this. The magic sum is 369 before squaring and 20,049 after squaring.

28	13	9	59	66	79	51	44	20
50	8	19	81	58	65	43	30	15
11	77	70	42	46	35	4	27	57
75	33	53	22	2	18	68	61	37
6	72	56	34	41	48	26	10	76
45	21	14	64	80	60	29	49	7
25	55	78	47	36	40	12	5	71
67	52	39	17	24	1	63	74	32
62	38	31	3	16	23	73	69	54

Bimagic square of order 9

Five Magic Squares Beyond Imagination

Nineteen Magic Squares in One

Figure 3.31 shows a complicated magic square, with beautifully overlapped subsquares, created by John Hendricks.[76] The overall 12×12 magic square has a magic sum of 870. Inside are four pandiagonal magic squares of order 8 with a magic sum of 580. There are nine pandiagonal magic squares of order 4 with a magic sum of 290.

Another Overlapping Square

Just as awesome is the magic square in Figure 3.32 by John Hendricks.[77] The lines indicate the positions of various internal subsquares.

94	57	76	63	142	9	124	15	118	33	100	39
75	64	93	58	123	16	141	10	99	40	117	34
69	82	51	88	21	130	3	136	45	106	27	112
52	87	70	81	4	135	33	129	28	111	46	105
96	59	74	61	144	11	122	13	120	35	98	37
73	62	95	60	121	14	143	12	97	38	119	36
71	84	49	86	23	132	1	134	47	108	25	110
50	85	72	83	2	133	24	131	26	109	48	107
92	55	78	65	140	7	126	17	116	31	102	41
77	66	91	56	125	18	139	8	101	42	115	32
67	80	53	90	19	128	5	138	43	104	29	114
54	89	68	79	6	137	20	127	30	113	44	103

Figure 3.31
Nineteen magic squares in one.

209	58	4	181	149	118	64	121	60	207	186	2	146	63	127
184	1	59	208	124	61	119	148	3	142	75	132	51	197	191
46	199	193	14	106	139	133	74	122	56	198	187	15	147	66
13	194	196	49	73	134	136	109	192	6	137	71	123	52	210
205	54	8	185	145	114	68	125	67	135	57	201	182	11	138
188	5	55	204	128	65	115	144	206	183	7	150	72	126	47
50	203	189	10	110	143	129	70	141	62	131	48	202	195	12
9	190	200	53	69	130	140	113	86	96	157	173	26	36	217
214	31	24	178	95	164	85	156	97	83	116	216	37	23	176
179	100	154	76	219	43	20	82	111	161	98	22	171	221	38
88	215	44	25	169	91	159	101	158	112	81	41	218	172	21
16	174	103	155	89	220	34	117	90	92	153	177	30	32	213
160	79	211	39	28	170	104	152	93	87	120	212	33	27	180
35	29	175	94	151	84	223	78	107	165	102	18	167	225	42
99	163	80	224	40	19	166	105	162	108	77	45	222	168	17

Figure 3.32
More overlapping squares.

Four Diamonds

Figure 3.33 is a magic square by John Hendricks with four internal diamonds that are magic.[78]

An Inlaid Magic Square of Order 24

In 1991, John Hendricks created a magic square suitable for Zeus himself (Fig. 3.34).[79] The entire square contains all the numbers from 1 to 576 and sums to 6924 in each row, column, and diagonal. At upper left is an inlaid set of bordered squares surrounding a magic diamond. At upper right is a pandiagonal overlapping set of squares. At lower left is a pandiagonal magic square of order 9,

77	106	104	147	134	100	95	175	155	6	49	36	149	46
78	107	118	130	125	135	70	29	156	20	179	27	37	168
94	136	115	138	117	109	54	45	38	164	187	166	11	5
56	124	132	123	114	122	92	7	173	34	172	16	171	43
64	137	129	108	131	110	84	15	39	178	157	180	12	35
96	111	121	116	128	139	52	47	160	23	165	30	188	3
53	140	142	99	112	146	71	4	189	44	1	14	195	169
102	91	93	50	63	97	120	151	42	191	148	161	48	22
145	62	72	67	79	90	101	194	13	170	18	177	41	150
113	88	80	59	82	61	133	162	186	31	10	33	159	182
105	75	83	74	65	73	141	154	26	181	25	163	24	190
143	87	66	89	68	60	103	192	185	17	40	19	158	152
127	58	69	81	76	86	119	176	9	167	32	174	184	21
126	57	55	98	85	51	144	28	8	153	196	183	2	193

Figure 3.33
Four diamonds.

once thought to be impossible to construct. At bottom right is a set of bordered magic squares starting with an order-9 square.

Bimagic Inlay

Figure 3.35 is a twelfth-order standard magic square by John Hendricks with a sum of 870 that contains an inlaid bimagic square of order 8.[80] The inlay sums to 580 in the first degree and is pandiagonal. The inlay sums to 48,140 in the second degree and is standard.

```
576  14  22  17  19  15   9   5   7   2  12   4 573 565 575 570 572 568 562 558 560 555  11 553
313 251 243 248 246 250 256 260 258 263 253 261 316 324 314 319 317 321 327 331 329 334 254 336
505  59 304 383 234 236 370 222 224 263 373  69 525 510  40 524  34 198 352 445 274 116 515  49
385 179 544 308  54 143 476 534 123 469  58 189 405 164 394 150 400 442 284 102 208 349 395 169
433 131 418 414  51 191 421 311 531 188 184 141 453  30 520  44 514 112 205 346 452 270 443 121
337 227 466 464 536 176 546 186  61 138 136 237 357 404 154 390 160 356 438 280 109 202 347 217
193 371 376 140 360 431 301 171 296 462 226 381 213 512 147 407  42 277 106 212 342 448 203 361
 97 467 130 174 541 416  56 426  66 428 472 477 117 402  37 272 507 167 210 339 200 359 107 457
145 419 178 474  71 291 181 411 551 128 424 429 165 267 527 162 397  32 440 119 450  99 155 409
 25 539  64 133 548 459 126  68 479 294 538 549  45 157 392  27 287 522 354 195 344 215  35 529
265 299 229 219 368 366 232 380 378 239 298 309 285  47 282 517 152 387 104 455 114 435 275 289
 73 491 483 488 486 490 496 500 498 503 493 501  93  85  95  90  92  88  82  78  80  75  83 481
504  86 502 497 499 495 489 485 487 482 492 484  76  84  74  79  77  81  87  97  89  94 494  96
312 278 478 298 533 382 417 125  70 177 221 292 268 273 194 343 341 207 355 353 214 204 302 288
552  38 377 415 132  65 175 228 473 295 540 532  28  33 269 523 434 101  43 454 108 519 542  48
432 158  63 187 218 471 307 530 375 427 122 412 148 159 163 271 386 161 406 156 389 393 422 168
480 110 461 310 537 365 430 129  53 190 225 460 100 111 113 521 286  26 516  31 439 441 470 120
384 206 372 425 127  60 185 223 468 305 535 364 196 201 437 151 506 276  46 401 115 351 374 216
240 350  50 183 235 458 303 547 362 423 139 220 340 447 403  41  36 526 266 511 149 105 230 360
144 446 465 293 550 369 413 142  57 173 238 124 436 399  13 396 166 391 146 281 449 153 134 456
192 398 367 420 137  55 180 233 463 300 545 172 388 513  44  29 118 451 509  98 283  39 182 408
 72 518  67 170 231 475 290 543 379 410 135  52 508 348 358 209 211 345 197 199 338 279  62 528
241 232 262 257 259 255 249 245 247 242 252 244 333 325 335 330 332 328 322 318 320 315 326 364
 24 566   3   8   6  10  16  20  18  23  13  21 556 564 554 559 557 561 567 571 569 574 563   1
```

Figure 3.34

An inlaid magic square of order 24.

1	143	25	109	72	84	85	49	108	48	134	12
132	14	16	110	71	83	86	50	107	47	23	121
3	15	116	54	45	75	32	90	105	63	142	130
10	22	46	76	115	53	106	64	31	89	135	123
138	126	79	41	58	112	67	101	94	28	7	19
139	127	57	111	80	42	93	27	68	102	6	18
8	20	113	55	40	82	29	91	100	70	137	125
5	17	39	81	114	56	99	69	30	92	140	128
141	129	78	44	51	117	66	104	87	33	4	16
136	124	52	118	77	43	88	34	65	103	9	21
24	122	119	35	74	62	59	95	38	98	131	13
133	11	120	36	73	61	60	96	37	97	2	144

Figure 3.35

Bimagic inlay.

David Collison's Magic Partition

David Collison partitioned the fourteenth-order magic square in Figure 3.36 into four squares, four elbows, four Ts, and one cross.[81]

To get a feel for the odd magic pieces, consider the L shape in the upper right. We call it magic because the sums along the longest directions are 394 $(154 + 155 + 41 + 44,\ 42 + 43 + 156 + 153,\ 154 + 42 + 40 + 158,$ and $155 + 43 + 157 + 39)$ and the sums along the short directions are 197 $(44 + 153,\ 158 + 39,\ \text{etc.})$. Similarly, the other pieces may also be considered magic.

Another exquisite example by Collison is shown in Figure 3.37. This magic square of order 10 has a magic sum of 505 and is partitioned into four magic elbows and a magic cross.

154	155	41	44	2	6	190	192	8	193	38	35	161	160
42	43	156	153	195	191	5	7	189	4	159	162	37	36
40	157	91	105	104	94	3	194	83	113	112	86	163	34
158	39	102	96	97	99	196	1	110	88	89	107	33	164
177	20	98	100	101	95	140	57	90	108	109	87	171	26
24	173	103	93	92	106	58	139	111	85	84	114	167	30
176	23	178	17	137	136	59	131	65	63	172	27	29	166
174	21	19	180	60	61	66	138	132	134	25	170	31	168
22	175	75	121	120	78	135	62	67	129	128	70	165	32
18	179	118	80	81	115	133	64	126	72	73	123	28	169
146	51	82	116	117	79	188	9	74	124	125	71	45	152
52	145	119	77	76	122	11	186	127	69	68	130	151	46
54	55	144	141	187	183	13	15	181	12	147	150	49	48
142	143	53	56	10	14	182	184	16	185	50	47	149	148

Figure 3.36
Fourteenth-order magic square partitioned into four magic squares, four T-shapes, four elbows, and one cross.

39	38	64	61	35	65	69	33	82	19
62	63	37	40	66	36	68	32	83	18
78	79	24	21	100	1	34	67	17	84
22	23	77	80	2	99	31	70	20	81
26	75	97	96	3	91	9	7	86	15
76	25	4	5	10	98	92	94	16	85
30	71	41	60	95	6	11	14	88	89
27	74	44	57	93	8	90	87	12	13
73	28	59	43	55	45	51	54	48	49
72	29	58	42	46	56	50	47	53	52

Figure 3.37
Another magic partition

Order-15 Composite

The following order-15 magic square consists of nine order-5 magic squares, each with an order-3 inlaid diamond magic square.[82]

As is common with composite magic squares, the order-5 magic sums themselves form an order-3 magic square with the constant 1695. The magic constants of the order-3 magic diamonds form an order-3 magic square with the constant 1017.

560	645	490
495	565	635
384	291	342

Magic square of order-5
magic constants

336	387	294
297	339	381
384	291	342

Magic square of order-3
diamond constants

16	73	157	118	196	33	90	174	135	213	2	59	143	104	182
199	64	205	70	22	216	81	222	87	39	185	50	191	56	8
115	163	112	61	109	132	180	129	78	126	101	149	98	47	95
202	154	19	160	25	219	171	36	177	42	188	140	5	146	11
28	106	67	151	208	45	123	84	168	225	14	92	53	137	194
3	60	144	105	183	17	74	158	119	197	31	88	172	133	211
186	51	192	57	9	200	65	206	71	23	214	79	220	85	37
102	150	99	48	96	116	164	113	62	110	130	17	127	76	124
189	141	6	147	12	203	155	20	161	26	217	169	34	175	40
15	93	54	138	195	29	107	68	152	209	43	121	82	166	223
32	89	173	134	212	1	58	142	103	181	18	75	159	120	198
215	80	22	86	38	184	49	19	55	7	201	66	207	72	24
131	179	128	77	125	100	148	97	46	94	117	16	114	63	111
218	170	35	176	41	187	139	4	145	10	204	155	21	162	27
44	122	83	167	224	13	91	52	136	193	30	108	69	153	210

Order-15 composite

Super Overlapping Fifteenth-Order Magic Square

On page 252 is a superb fifteenth-order magic square (magic constant 1695) containing several overlapping smaller magic squares.[83] Two 6 × 6 magic squares are at the upper left and bottom right. Two overlapping 9 × 9 magic squares are at the upper right and bottom left. These have a 3 × 3 magic square in common at the center of the figure. If this is not sufficiently impressive, consider the overlapping 5 × 5 magic squares, the overlapping 7 × 7 squares, the central 11 × 11 magic square that overlaps all the others, and the numerous other squares awaiting your discovery.

Isn't this a tasty meal for a king?

225	216	3	222	5	7	73	143	75	141	77	139	79	152	138
10	1	223	4	221	219	153	83	151	85	149	87	147	88	74
6	220	11	18	212	211	89	129	91	127	93	136	126	81	145
218	8	213	210	12	17	137	97	135	99	133	100	90	82	144
2	224	14	15	215	208	101	119	103	124	118	95	181	150	76
217	9	214	209	13	16	125	107	123	108	102	96	130	84	142
77	149	71	155	69	157	112	117	110	105	121	134	92	148	78
52	174	64	162	70	156	111	113	115	106	120	98	128	86	140
181	45	180	46	186	40	116	109	114	122	104	132	94	146	80
53	173	66	160	168	154	37	167	39	29	36	194	193	24	202
178	48	163	63	72	58	189	59	187	195	192	30	35	20	206
55	171	169	158	38	161	44	159	62	32	33	197	190	200	26
176	50	68	57	188	65	182	67	164	136	191	31	34	199	27
184	165	41	172	43	170	47	146	49	21	204	23	25	207	198
61	42	185	54	183	56	179	80	177	205	22	203	201	28	19

Super overlapping fifteenth-order magic square

Super Overlapping Thirteenth-Order Magic Square

The following 13×13 magic square is actually a combination of thirteen different-size overlapping magic squares.[84] Let's ask the Metropolitan Museum of Art in New York City to hang this awesome creation in their galleries!

Here are all the magic squares I could find. Perhaps you can find more.

- *A*–4 × 4 magic square at upper left. All rows, columns, and the two main diagonals sum to 340. (To help you locate this square, the four corners are 157, 147, 22, and 14.) Example sum: $157 + 13 + 23 + 147 = 340$. Also, seven sets of 2×2 square arrays also sum to this magic constant. Example sum: $157 + 13 + 145 + 25 = 340$.

157	13	23	147	109	31	111	138	36	66	102	100	72
145	25	17	153	61	139	59	32	134	104	68	98	70
16	154	144	26	57	56	30	112	136	99	105	60	110
22	148	156	14	113	114	140	58	34	65	71	133	37
97	73	94	76	151	18	21	89	146	135	35	29	141
79	91	78	92	27	82	150	155	11	63	107	33	137
74	96	75	95	143	159	15	20	88	115	55	101	69
90	80	93	77	19	24	81	149	152	54	116	103	67
164	6	3	167	85	142	158	12	28	64	106	108	62
7	163	168	86	1	132	44	39	125	50	48	118	124
162	8	84	2	169	38	126	131	45	120	122	52	46
5	83	161	10	166	129	43	40	128	123	117	49	51
87	165	9	160	4	41	127	130	42	47	53	121	119

Super overlapping thirteenth-order magic square

- B–4×4 magic square directly beneath A and with the same properties. (The four corners are 97, 76, 90, and 77.) Example sum: $97 + 73 + 94 + 76 = 340$.
- C–5×5 central magic square. All rows, columns, and the two main diagonals sum to 425. (The four corners are 151, 146, 85, and 28.) Example sum: $151 + 18 + 21 + 89 + 146 = 425$.
- D–4×4 magic square at bottom center; same properties as A. (The four corners are 132, 125, 41, and 42.)
- E–4×4 magic square at bottom right; same properties as D. (The four corners are 50, 124, 47, and 119.)
- F–9×9 magic square at bottom left in gray. All rows, columns, and the two main diagonals sum to 765. (The four corners are 97, 146, 87, and 42.)
- G–5×5 magic square at bottom left; same properties as C. (The four corners are 164, 85, 87, and 4.)

- H–9 × 9 magic square at upper right; same properties as F. (The four corners are 109, 72, 85, and 62.)
- I–7 × 7 magic square roughly in the center. All rows, columns, and the two main diagonals sum to 595. (The four corners are 57, 105, 85, and 106.)
- J–13 × 13 magic square (the entire array). All rows, columns, and the two main diagonals sum to 1105.
- K–11 × 11 magic square at upper right. All rows, columns, and the two main diagonals sum to 935. (The four corners are 23, 72, 84, and 46.)
- L–7 × 7 magic square, roughly in center. Same properties as I. (The four corners are 94, 146, 9, and 42.)
- M–3 × 3 semimagic square. All rows, columns, but (sadly) only one main diagonal sum to 255. (The four corners are 3, 85, 84, and 169.)

All 169 cells in this array are included in at least two different magic squares. The cell containing the number 85 appears in six different magic squares. It is also in the correct diagonal of the 3 × 3 semimagic square! This array was invented by L. S. Frierson and appears in *Magic Squares & Cubes* by W. S. Andrews, which was published in 1917.

1750 Bimagic Square

We've discussed bimagic or doubly magic squares a few times in this book. It is possible to construct magic squares of order 8 or greater that are doubly magic; that is, if the number in each cell is replaced by its square, the resulting square is also magic. I enjoy the square[85] on page 255 in particular because it was supposedly constructed in 1750, the year Johann Sebastian Bach died, which is the first two numbers in the square.[86]

An Order-9 Pandiagonal Magic Square

Until the 1990s, everyone believed that it was impossible to construct a regular pandiagonal order-9 magic square–partly because there is no very obvious recipe (or pattern) to use to create a regular 9 × 9 square. However, in 1996, Japanese magic square expert Mr. Gakuho Abe discovered a whole series of such squares.[87]

17	50	43	4	32	75	66	21
31	76	65	22	14	53	40	7
0	47	54	13	25	62	71	36
26	61	72	35	3	44	57	10
45	2	11	56	60	27	34	73
63	24	37	70	46	1	12	55
52	15	6	41	77	30	23	64
74	33	20	67	51	16	5	42

1750 bimagic square

1	16	5	30	45	77	56	71	22
41	47	61	67	73	9	15	21	35
69	75	8	14	20	34	40	46	63
13	19	36	42	48	62	68	74	7
53	32	64	79	58	12	27	6	38
81	60	11	26	5	37	52	31	66
25	4	39	54	33	65	80	59	10
29	44	76	55	70	24	3	18	50
57	72	23	2	17	49	28	43	78

An order-9 pandiagonal magic square

The square above is composed of consecutive numbers from 1 to 81. All rows, columns, and the two main diagonals sum to the constant 369. Being pandiagonal, the broken diagonals also sum to the magic constant, and the square can be transposed to another by moving any column or row to the opposite side.

Pearl Harbor

The order-11 magic square in this section consists of a normal order-9 composite magic square divided into its nine order-3 magic squares with forty additional 41s. The magic constant is 11×41. Mr. E. W. Shineman constructed this square to commemorate the year 1941, the year the Japanese attacked Pearl Harbor![88] The magic constants of each of the nine order-3 magic squares themselves form an order-3 magic square, a feature of most composite magic squares. Try constructing similar magic squares with numbers other than 41.

71	64	69	41	8	1	6	41	53	46	51
66	78	70	41	3	5	7	41	48	50	52
67	72	65	41	4	9	2	41	49	54	47
41	41	41	41	41	41	41	41	41	41	41
26	19	24	41	44	37	42	41	62	55	60
21	23	25	41	39	41	43	41	57	59	61
22	27	20	41	40	45	38	41	58	63	56
41	41	41	41	41	41	41	41	41	41	41
35	28	33	41	80	73	78	41	17	10	15
30	32	34	41	75	77	79	41	12	14	16
31	36	29	41	76	81	74	41	13	18	11

Pearl Harbor magic square

Prime Number Heterosquares

In September 1998, Carlos Rivera of Monterrey, Nuevo León, México, discovered two unusual heterosquares.[89] The order-3 heterosquare on the left consists of nine prime numbers. The three rows, three columns, and two main diagonals all sum to different prime numbers. (The sums are shown at the sides of the order-3 square.) The sum of all nine cells is also a prime number, shown in bold. Does a square exist with a smaller total using nine unique primes?

The square on the right has identical features but in addition consists of *consecutive* primes. Does a square exist with a smaller total or a square that contains nine consecutive primes?

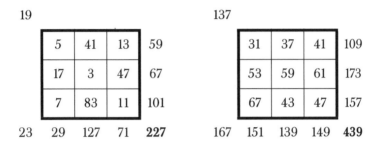

Two prime number heterosquares

Pandiagonal with Inlaid Letters

Displayed on page 258 is an order-16 pandiagonal pure magic square that was constructed in September 1998 by E. W. Shineman for Harvey Heinz.[90] It uses the consecutive numbers from 1 to 256. Each of the sixteen rows, columns, and diagonals sums to the constant 2056. The cells in "E. S." also sum to 2056 and the cells in "H. H." sum to 2056×2. Notice the 98 in the upper cell of the left column signifying the year in which the square was constructed.

Patterns in a Fifth-Order Square

Throughout our journey with magic squares, we've seen lots of different ways for producing sums in addition to summing the rows, columns, and diagonals. Here are some additional patterns you may not have considered.[91] Consider the order-5 square on the bottom of page 258.

98	79	178	95	162	63	194	47	210	255	2	239	18	143	114	159
158	179	78	163	94	195	62	211	46	3	254	19	238	115	142	99
100	77	180	93	164	61	196	45	212	253	4	237	20	141	116	157
155	182	75	166	91	198	59	214	43	6	251	22	235	118	139	102
101	76	181	92	165	60	197	44	213	252	5	236	21	140	117	156
153	184	73	168	89	200	57	216	41	8	249	24	233	120	137	104
103	74	183	90	167	58	199	42	215	250	7	234	23	138	119	154
151	186	71	170	87	202	55	218	39	10	247	26	231	122	135	106
105	72	185	88	169	56	201	40	217	248	9	232	25	136	121	152
149	188	69	172	85	204	53	220	37	12	245	28	229	124	133	108
107	70	187	86	171	54	203	38	219	246	11	230	27	134	123	150
148	189	68	173	84	205	52	221	36	13	244	29	228	125	132	109
110	67	190	83	174	51	206	35	222	243	14	227	30	131	126	147
146	191	66	175	82	207	50	223	34	15	242	31	226	127	130	111
112	65	192	81	176	49	208	33	224	241	16	225	32	129	128	145
160	177	80	161	96	193	64	209	48	1	256	17	240	113	144	97

Pandiagonal with inlaid letters

1	22	18	14	10
19	15	6	2	23
7	3	24	20	11
25	16	12	8	4
13	9	5	21	17

Order-5 square with hidden patterns

In addition to the traditional sums, the following patterns sum to 65.

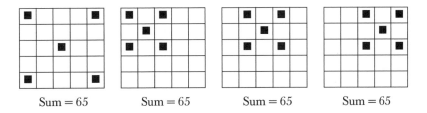

Sum = 65   Sum = 65   Sum = 65   Sum = 65

The following are some additional patterns that produce the magic sum. Here the ⚡ symbol indicates that the corresponding number is to be *subtracted* rather than added. These patterns can be reflected and moved in various ways. I like the look of the "stealth fighter" configurations. Imagine gliding on one of these fast crafts over an infinite mathematical landscape . . .

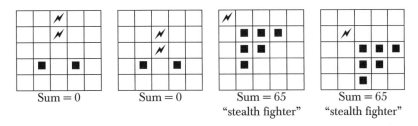

Sum = 0   Sum = 0   Sum = 65   Sum = 65
"stealth fighter"   "stealth fighter"

Derksen Order-25 Square

The following 25×25 magic square,[92] by Harm Derksen of The Institute of Mathematics at the University of Basel, Switzerland, has standard and special properties:

- The entries of the square are consecutive numbers from 1 to $625 = 25^2$.
- All the column, row, and diagonal sums are $7825 = 25 \times (1 + 625)/2$.
- If the square is divided into twenty-five 5×5 squares, these subsquares are magic too; the row, column, and diagonal sums are $1565 = 5 \times (1 + 625)/2$.
- If one squares all the number entries, the square remains magic; all row, column, and diagonal sums are equal to 3,263,025.

We're sure there are additional symmetries waiting to be found. Harm used Maple, a mathematics computer program, to generate this large magic square.

1	443	235	547	339	283	100	387	179	616	565	352	44	456	148	217	509	321	113	405	499	161	578	270	57
157	599	261	53	495	439	226	543	335	22	91	383	200	612	279	373	40	452	144	556	505	317	109	421	213
313	105	417	209	521	595	257	74	486	153	247	539	326	18	435	379	191	608	300	87	31	473	140	552	369
469	131	573	365	27	121	413	205	517	309	253	70	482	174	586	535	347	14	426	243	187	604	291	83	400
625	287	79	391	183	127	569	356	48	465	409	221	513	305	117	61	478	170	582	274	343	10	447	239	526
587	254	66	483	175	244	531	348	15	427	396	188	605	292	84	28	470	132	574	361	310	122	414	201	518
118	410	222	514	301	275	62	479	166	583	527	344	6	448	240	184	621	288	80	392	461	128	570	357	49
149	561	353	45	457	401	218	510	322	114	58	500	162	579	266	340	2	444	231	548	617	284	96	388	180
280	92	384	196	613	557	374	36	453	145	214	501	318	110	422	491	158	600	262	54	23	440	227	544	331
431	248	540	327	19	88	380	192	609	296	370	32	474	136	553	522	314	101	418	210	154	591	258	75	487
423	215	502	319	106	55	492	159	596	263	332	24	436	228	545	614	276	93	385	197	141	558	375	37	454
554	366	33	475	137	206	523	315	102	419	488	155	592	259	71	20	432	249	536	328	297	89	376	193	610
85	397	189	601	293	362	29	466	133	575	519	306	123	415	202	171	588	255	67	484	428	245	532	349	11
236	528	345	7	449	393	185	622	289	76	50	462	129	566	358	302	119	406	223	515	584	271	63	480	167
267	59	496	163	580	549	336	3	445	232	176	618	285	97	389	458	150	562	354	41	115	402	219	506	323
359	46	463	130	567	511	303	120	407	224	168	585	272	64	476	450	237	529	341	8	77	394	181	623	290
390	177	619	281	98	42	459	146	563	355	324	111	403	220	507	576	268	60	497	164	233	550	337	4	441
541	333	25	437	229	198	615	277	94	381	455	142	559	371	38	107	424	211	503	320	264	51	493	160	597
72	489	151	593	260	329	16	433	250	537	606	298	90	377	194	138	555	367	34	471	420	207	524	311	103
203	520	307	124	411	485	172	589	251	68	12	429	241	533	350	294	81	398	190	602	571	363	30	467	134
195	607	299	86	378	472	139	551	368	35	104	416	208	525	312	256	73	490	152	594	538	330	17	434	246
346	13	430	242	534	603	295	82	399	186	135	572	364	26	468	412	204	516	308	125	69	481	173	590	252
477	169	581	273	65	9	446	238	530	342	286	78	395	182	624	568	360	47	464	126	225	512	304	116	408
508	325	112	404	216	165	577	269	56	498	442	234	546	338	5	99	386	178	620	282	351	43	460	147	564
39	451	143	560	372	316	108	425	212	504	598	265	52	494	156	230	542	334	21	438	382	199	611	278	95

Derksen order-25 square

Tilegrams

Dave Harper from Toronto, Ontario, has found beautiful, hidden patterns that underlie the very essence of magic squares. These patterns are made clear to the human eye when the numbers of the square are replaced by symmetric symbols.[93]

We can use computer graphics like a stain applied to wood to reveal the hidden wood grains. As you will see, the attractiveness of the resulting patterns result from (1) the choice of a highly symmetric set of symbols to represent the numbers of the magic square and (2) the inherent symmetries of the magic square that are required to produce the square's magic numeric properties.

The remainder of this section deals with patterns representing continuous magic squares (also called Nasik or pandiagonal magic squares) of order 4 and various sets of symbols that can be used to highlight the magic structures. I look forward to hearing from readers who have applied this approach to larger squares using different symbols.

As a review and reminder from chapter 2, in a simple magic square all of the columns, rows, and diagonals add up to the same value. A Nasik magic square has the additional property that all broken diagonals also add up to that same number. Examples of broken diagonals from the square below are 2, 1, 13, and 14 and 8, 14, 7, and 1.

8	5	2	15
6	11	12	1
13	0	7	10
3	14	9	4

Nasik square

In this book, we've usually started the magic square with the number 1. However, for the patterns in this section, the sequence of numbers from 0 to 15 are used. This means that the magic sum is 30 rather than the usual 34 when the sequence of numbers from 1 to 16 is used.

There are eight broken diagonals in an order-4 Nasik square. These diagonals become "continuous" if you replicate one of these squares in all directions to create an orthogonal tiling of magic squares. Any order-4 square in this numbered tiling of the plane will be magic. Four such squares are shown on page 262. The partial tiling generated from our example order-4 square shows that any four consecutive numbers (diagonal, horizontal, or vertical) add up to 30.

Here are three different numbering schemes for continuous Nasik squares of order 4:

8	5	2	15
6	11	12	1
13	0	7	10
3	14	9	4

Nasik square A

7	10	4	9
12	1	15	2
11	6	8	5
0	13	3	14

Nasik square B

7	9	14	0
12	2	5	11
1	15	8	6
10	4	3	13

Nasik square C

8	5	2	15	8	5	2	15	8	5	2	15
6	11	12	1	6	11	12	1	6	11	12	1
13	0	7	10	13	0	7	10	13	0	7	10
3	14	9	4	3	14	9	4	3	14	9	4
8	5	2	15	8	5	2	15	8	5	2	15
6	11	12	1	6	11	12	1	6	11	12	1
13	0	7	10	13	0	7	10	13	0	7	10
3	14	9	4	3	14	9	4	3	14	9	4
8	5	2	15	8	5	2	15	8	5	2	15
6	11	12	1	6	11	12	1	6	11	12	1
13	0	7	10	13	0	7	10	13	0	7	10
3	14	9	4	3	14	9	4	3	14	9	4

Nasik continuous array

All other continuous Nasik squares of order 4 can be generated from these three basic number motifs. For example, if you replicate one of these squares to form a continuous number tiling, then there are sixteen different ways you can create a 4 × 4 subset of the resulting tiling. To produce these sixteen different squares, each square has a different number in the top left corner. Square A, B, or C can form a continuous magic square. This means there are 3 × 16 = 48 variations based on tiling subsets. You can also use any one of the tiling subsets and generate eight more continuous magic squares by 90 degree rotations and reflections. This gives a total of 8 × 48 = 384 Nasik squares of order 4 generated from the original three basic number layouts.

Various mappings from numbers in the magic square to tiling patterns can be used to reveal a limitless world of beauty and intricacy secreted within magic squares. Dave Harper suggests a mapping that represents the sequence of numbers from 0 to 15 as "pictures" of their four-digit binary numbers. (For background on binary numbers, see note 3 for this chapter.) In all cases, a square is

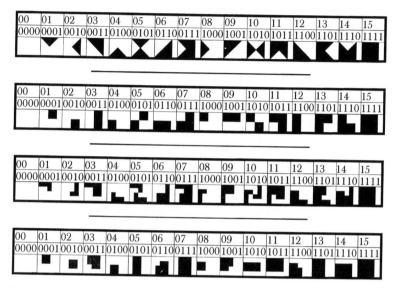

Figure 3.38
Harper mapping schemes.

symmetrically divided up into four equal segments that represent
one of the four digits in a binary number from 0000 to 1111 (0 to
15). A segment is colored white to represent 0 and black to represent 1. Four suggested mapping schemes are shown in Figure 3.38.
The first scheme seems to produce the most consistently pleasing
pictures due to the scheme's high degree of symmetry.

Note that for each of these schemes, rotating the small picture
(corresponding to a cell) 90 degrees is equivalent to multiplying
the number it represents by 2 (and, if necessary, changing the
result back to the range 0 to 15 by subtracting 15).

Figures 3.39 through 3.44 show a sampling of patterns created
by mapping various Nasik squares to patterns using the first mapping scheme in Figure 3.38. There are many open areas for future
research and art. For example, the numbering schemes used here
are useful for fourth-order magic squares because they represent
the numbers 0 through 15 as a set of symmetrical symbols. What

sets of symbols could be used for larger magic squares? How many different patterns exist for Nasik magic squares? What do the beautiful symmetries tell us about the numerical structure of the square?

Figure 3.39
Magic tiling for square A.

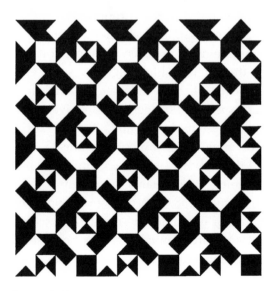

Figure 3.40
Magic tiling for square A rotated 90 degrees.

Figure 3.41
Magic tiling for square A rotated 180 degrees.

Figure 3.42
*Magic tiling for square A after vertical mirror reflection
and then rotation by 90 degrees.*

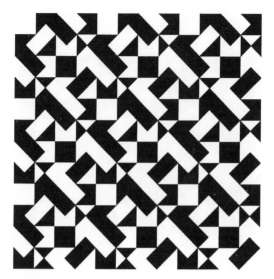

Figure 3.43
Magic tiling for square B after vertical mirror reflection and then rotation by 90 degrees.

Figure 3.44
Magic tiling for square B after rotation by 270 degrees.

Physical Magic Square World Records

Ralf Laue and his "Rekord-Klub Saxonia" continue to track world records concerning physical embodiments of magic squares. For a square to be considered, it must conform to certain rigid criteria:[94]

1. The magic square must be written or printed on paper. It is not sufficient just to calculate it by a computer.
2. Magic squares can be created from many sheets of paper, but they must be positioned together to form the array of numbers. This physical embodiment must be in the shape of a square, not a rectangle.
3. To verify the sums in each row, column, and diagonal are correct, the program used to generate the numbers must be run while under supervision of a computer/mathematics specialist who can verify that the program is correct.

Here are the world records that have been verified since 1975:

1.	105×105	Richard Suntag (Pomona, USA)	1975
2.	501×501	Gerolf Lenz (Wuppertal, Germany)	1979
3.	897×897	Frank Tast and Uli Schmidt (Pforzheim, Germany)	1987
4.	1000×1000	Christian Schaller (Munich, Germany)	1988
5.	2001×2001	Sven Paulus, Ralph Bülling, Jörg Sutter (Pforzheim, Germany)	1989
6.	2121×2121	Ralf Laue (Leipzig, Germany)	1991
7.	3001×3001	Louis Caya (Sainte-Foy, Canada)	1994
8.	Largest magic square written by hand: 1111×1111	Norbert Behnke (Krefeld, Germany)	1990

Grand-Prize Winner: Hendricks's Ornate Inlaid Magic Cube!

Many times throughout this book, I've mentioned magic square guru John Hendricks. Before telling you about his magic object to which I've awarded the grand prize, let me digress and tell you more about this man's singular fascination with magic squares.

For 33 years, John Hendricks worked for the Canadian Meteorological Service, and he retired in 1984. Throughout his career, he was also known for his contributions to climatological statistics. After retirement, John gave many public lectures on magic squares, and he developed a course on magic squares and cubes for mathematically inclined students.

John Hendricks started collecting magic squares and cubes when he was thirteen years old. This became a hobby and sometimes even an obsession. He never dreamed he would one day make startling contributions to the field, becoming the first person in the world to successfully construct four-, five-, and six-dimensional models of magic hypercubes and write prolifically on the subject in the *Journal of Recreational Mathematics.* John has also extended the knowledge of magic squares and cubes, especially the ornate and embedded varieties. His numerous books include *Magic Squares to Tesseracts by Computer, Inlaid Magic Squares and Cubes,* and *All Third Order Magic Tesseracts.*

Without further ado, here is my favorite from John Hendricks's ingenious collection of magic objects: a $12 \times 12 \times 12$ magic cube that houses eight *interchangeable* interior magic cubes that are each encased by "wrappers" of six pandiagonal magic squares.[95] Figure 3.45 is a schematic diagram showing the outer shell of the order-12 cube and positions of the inlaid magic cubes. We call the overall cube inlaid because it contains eight inlaid subcubes. We call the cube ornate because of all the special magic squares. Figures 3.46–3.58 show the various layers of the twelfth-order magic cube and inlaid pandiagonal magic squares of order 4. Figure 3.59 shows the locations of the pandiagonal wrapping squares and the pantriagonal magic cubes.

The large twelfth-order cube enjoys the standard requirement for all cubes that each row, each column, each pillar, and each of the four space-diagonals called *triagonals* sum to 10,374. The consecutive numbers from 1 to 12^3 are used.

Each of the eight inner subcubes are special pantriagonal cubes of order four that sum to 3458 in standard ways but also in all broken triagonals. Moreover, the eight inner subcubes are independent and can be rotated and/or reflected in their forty-eight aspects at will, and they are also completely interchangeable!

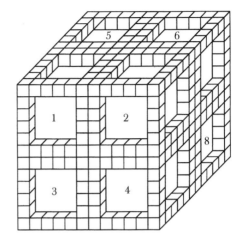

Figure 3.45
The fixed outer shell of John Hendricks's order-12 cube. This figure should help you visualize the positions of eight interchangeable inlaid magic squares and cubes. Cube number 7 is hidden from view in the back.

942	1230	355	222	1651	643	1075	67	1518	1363	510	798
966	474	1400	4011	1183	619	1110	174	1700	101	1483	763
751	1267	317	1340	534	1122	607	1543	41	1616	258	978
882	1328	546	1255	329	703	1026	1628	246	1555	29	847
859	389	1195	462	1412	1014	715	113	1471	186	1688	870
775	487	1386	1495	90	1098	666	1674	199	378	1207	919
811	1242	343	234	1639	1062	630	55	1530	1351	522	955
727	306	1280	569	1303	1146	583	6	1580	269	1603	1002
990	1435	437	1172	414	595	1134	1711	161	1448	138	739
835	1160	426	1423	449	1038	691	1460	126	1723	149	894
906	557	1315	294	1292	679	1050	281	1591	18	1568	823
930	499	1374	1507	78	655	1087	1662	211	366	1219	786

Figure 3.46
Layer 1 (top layer) of the order-12 cube in Figure 3.45 showing four inlaid pandiagonal magic squares of order 4. These four "panels," along with others not shown in this layer, form a coating around the actual internal magic cubes.

940	1245	352	1371	490	640	1089	1534	63	1660	201	789
760	563	1308	422	1165	904	681	157	1718	12	1571	1113
1125	289	1274	456	1439	691	892	143	1464	266	1585	748
610	1296	311	1417	343	1042	831	1442	121	1607	288	975
963	1310	565	1163	420	819	1054	1716	155	1573	14	622
772	1209	364	1359	526	1101	664	1498	75	1648	237	921
813	508	1353	382	1215	1060	633	219	1642	93	1504	952
735	1394	457	1271	336	879	706	1632	263	1465	98	1138
1150	1188	395	1333	542	718	867	1550	37	1691	180	723
585	397	1190	540	1331	1017	856	35	1548	182	1693	1000
988	479	1416	314	1249	844	1029	241	1610	120	1487	597
933	496	1389	346	1227	657	1072	207	1678	57	1516	796

Figure 3.47
Layer 2 of the order-12 cube in Figure 3.45. Here we begin to slice through the four subcubes and therefore see the inlaid top layers of the pantriagonal magic cubes of order 4.

790	483	1378	357	1240	1090	639	196	1665	70	1527	939
970	1429	446	1284	299	826	1047	1595	276	1454	133	615
603	1175	432	1298	553	1035	838	1561	2	1728	167	982
1120	410	1153	575	1320	688	897	24	1583	145	1706	753
765	444	1427	301	1286	909	676	278	1597	131	1452	1108
958	519	1366	369	1204	627	1066	232	1653	82	1491	807
915	1222	375	1348	513	670	1095	1509	88	1635	226	778
993	552	1343	385	1178	849	1024	170	1681	47	1560	592
580	326	1261	467	1404	1012	861	108	1475	253	1622	1005
1143	1259	324	1406	469	711	874	1477	110	1620	251	730
742	1321	530	1200	407	886	699	1703	192	1538	25	1131
795	1234	339	1384	501	1071	658	1521	52	1671	214	934

Figure 3.48
Layer 3 of the order-12 cube in Figure 3.45.

947	358	1239	484	1377	647	1082	69	1528	195	1666	782
1119	312	1295	433	1418	587	898	122	1441	287	1608	754
766	566	1309	419	1164	910	675	156	1715	13	1574	1107
969	1307	564	1166	421	825	1048	1717	158	1572	11	616
604	1273	290	1440	455	1036	837	1463	144	1586	265	981
779	370	1203	520	1365	1094	671	71	1492	231	1654	914
806	1347	514	1221	376	1067	626	1636	225	1510	87	959
1144	1189	398	1332	539	712	873	1547	36	1694	181	729
741	1415	480	1250	313	885	700	1609	242	1488	119	1132
994	458	1393	335	1272	850	1023	264	1631	97	1466	591
579	396	1187	541	1334	1011	862	38	1549	179	1692	1006
926	1383	502	1233	340	650	1079	1672	213	1522	51	803

Figure 3.49
Layer 4 of the order-12 cube in Figure 3.45.

781	1372	489	1246	351	1081	648	1659	202	1533	64	948
609	1154	409	1319	546	1041	832	1584	23	1705	146	976
964	1428	443	1285	302	820	1053	1598	277	1451	132	621
759	445	1430	300	1283	903	682	275	1596	134	1453	1114
1126	431	1176	554	1297	694	891	1	1562	168	1727	747
949	1360	525	1210	363	636	1057	1647	238	1497	76	816
924	381	1216	507	1354	661	1104	94	1503	220	1641	769
586	323	1260	470	1405	1018	855	109	1478	252	1619	999
987	529	1322	408	1199	843	1030	191	1704	26	1537	598
736	1344	551	1177	386	880	705	1682	169	1559	48	1137
1149	1262	325	1403	468	717	868	1476	107	1621	254	724
804	345	1228	495	1390	1080	649	58	1515	208	1677	925

Figure 3.50
Layer 5 of the order-12 cube in Figure 3.45.

```
785   497  1376 1505  80  1088 656 1664 209  368  1217 929
761   475  1397 404  1182 1112 620 175  1697 104  1482 965
980  1266  320  1337 535  605  1121 1542 44  1613 259  752
845  1325  547  1254 332  1028 704 1625 247 1554  32   881
872   392  1194 463  1409 713  1013 116  1470 187 1685  860
956  1244  341  236  1637 629  1061 53  1532 1349 524  812
920   488  1385 1496 89   665  1097 1673 200  377 1208  776
1004  307  1277 572  1302 581  1145  7  1577 272 1602  728
737  1434  440  1169 415  1136 596 1710 164 1445 139  989
896  1157  427  1422 452  689  1037 1457 127 1722 152  836
821   560  1314 295  1289 1052 680 284  1590 19  1565  905
797  1229  356  221  1652 1076 644 68  1517 1364 509  941
```

Figure 3.51
Layer 6 of the order-12 cube in Figure 3.45.

```
788  1220 365  212  1661 1085 653  77  1508 1373 500  932
824  1253 331  1326 548  1049 677 1553  31 1626 248  908
893   464 1410 391  1193 692  1040 188 1686 115 1469  833
740   403 1181 476  1398 1133 593 103  1481 176 1698  992
1001 1338 536  1265 319  584  1148 1614 260 1541  43   725
953   521 1352 1529 56   632  1064 1640 233 344 1241  809
917  1205 380  197  1676 668  1100 92  1493 1388 485  773
869  1421 451  1158 428  716  1016 1721 151 1458 128  857
848   296 1290 559  1313 1025 701  20  1566 283 1589  884
977   571 1301 308  1278 608  1124 271 1601  8  1578  749
764  1170 416  1433 439  1109 617 1446 140 1709 163  968
800   512 1361 1520 65  1073 641 1649 224 353 1232  944
```

Figure 3.52
Layer 7 of the order-12 cube in Figure 3.45.

```
 792 1391  494 1225  348 1092  637 1680  205 1514   59  937
 746 1161  418 1312  567  890  695 1575   16 1714  153 1127
1115 1419  436 1294  309  683  902 1605  286 1444  123  758
 624  454 1437  291 1276 1056  817  268 1587  141 1462  961
 973  424 1167  561 1306  829 1044   10 1569  159 1720  612
 960 1355  506 1213  384  625 1068 1644  217 1502   95  805
 913  362 1211  528 1357  672 1093   73 1500  239 1646  780
 721  316 1251  477 1414  865  720  118 1485  243 1612 1152
1140  538 1329  399 1192  708  877  184 1695   33 1546  733
 599 1335  544 1186  393 1031  842 1689  178 1552   39  986
 998 1269  334 1396  459  854 1019 1467  100 1630  261  587
 793  350 1247  492 1369 1069  660   61 1536  203 +658  936
```

Figure 3.53
Layer 8 of the order-12 cube in Figure 3.45.

```
 938  337 1236  503 1382  638 1091   50 1523  216 1669  791
 984  303 1288  442 1425  840 1033  129 1450  280 1599  601
 613  573 1318  412 1155 1045  828  147 1708   22 1581  972
1106 1300  555 1173  430  674  911 1726  165 1563    4  767
 755 1282  297 1431  448  899  686 1456  135 1593  274 1118
 770  373 1224  515 1346 1103  662   86 1511  228 1633  923
 815 1368  517 1202  371 1058  635 1655  230 1489   84  950
1007 1198  405 1323  532  863 1010 1540   27 1701  190  578
 590 1408  471 1257  322 1022  851 1618  249 1479  112  995
1129  465 1402  328 1263  697  888  255 1624  172 1683 1141
 732  387 1180  550 1341  876  709   45 1558  172 1683 1141
 935 1380  481 1238  359  659 1070 1667  194 1525   72  794
```

Figure 3.54
Layer 9 of the order-12 cube in Figure 3.45.

783	504	1381	338	1325	1083	646	215	1670	49	1524	946
1105	1438	453	1275	292	673	912	1588	267	1461	142	768
756	1168	423	1305	562	900	685	1570	9	1719	160	1117
983	417	1162	568	1311	839	1034	15	1576	154	1713	602
614	435	1420	310	1293	1046	827	285	1606	124	1443	971
951	516	1345	374	1223	634	1059	227	1634	85	1512	814
922	1201	372	1367	518	663	1102	1490	83	1656	229	771
1130	543	1336	394	1185	698	887	177	1690	40	1551	743
731	333	1270	460	1395	875	710	99	1468	262	1629	1142
1008	1252	315	1413	478	864	1009	1486	117	1611	244	577
589	1330	537	1191	400	1021	852	1696	183	1545	34	996
802	1237	360	1379	482	1078	651	1526	71	1668	193	927

Figure 3.55
Layer 10 of the order-12 cube in Figure 3.45.

945	1226	347	1392	493	645	1084	1513	60	1679	206	784
623	556	1299	429	1174	1055	818	166	1725	3	1564	962
974	298	1281	447	1432	830	1043	136	1455	273	1594	611
745	1287	304	1426	441	889	696	1449	130	1600	279	1128
1116	1317	574	1156	411	684	901	1707	148	1582	21	757
777	1214	383	1356	505	1096	669	1501	96	1643	218	916
808	527	1358	361	1212	1065	628	240	1645	74	1499	957
600	1401	466	1264	327	1032	841	1623	256	1474	105	985
997	1179	388	1342	549	853	1020	1557	46	1684	171	588
722	406	1197	531	1324	866	719	28	1539	189	1702	1151
1139	472	1407	321	1258	707	878	250	1617	111	1480	734
928	491	1370	349	1248	652	1077	204	1657	62	1535	801

Figure 3.56
Layer 11 of the order-12 cube in Figure 3.45.

943	511	1362	1519	66	642	1074	1650	223	354	1231	799
907	1256	330	1327	545	678	1051	1556	30	1627	245	822
834	461	1411	390	1196	1039	690	185	1687	114	1472	895
991	402	1184	473	1399	594	1135	102	1484	173	1699	738
726	1339	533	1268	318	1174	582	1615	257	1544	42	1003
774	1206	379	198	1675	1099	667	91	1494	1387	486	918
810	523	1350	1531	54	1063	531	1638	235	342	1243	954
858	1424	450	1159	425	1015	714	1724	150	1459	125	871
883	293	1291	558	1316	702	1027	17	1567	282	1592	846
750	570	1304	305	1279	1123	606	270	1604	5	1579	979
967	1171	413	1436	438	618	1111	1447	137	1712	162	762
931	1218	367	210	1663	654	1086	79	1506	1375	498	787

Figure 3.57
Layer 12 of the order-12 cube in Figure 3.45. In this slice we see four more panels that surround the magic cubes in layers 2 through 11.

930	499	1374	1507	78	655	1087	1662	211	366	1219	786
933	496	1389	346	1227	657	1072	207	1678	57	1516	796
795	1234	339	1384	501	1071	658	1521	52	1671	214	934
926	1383	502	1233	340	650	1079	1672	213	1522	51	803
804	345	1228	495	1390	1080	649	58	1515	208	1677	925
797	1229	356	221	1652	1076	644	68	1517	1364	509	941
800	512	1361	1520	65	1073	641	1649	224	353	1232	944
793	350	1247	492	1369	1069	660	61	1536	203	1658	936
935	1380	481	1238	359	659	1070	1667	194	1525	72	794
802	1237	360	1379	482	1078	651	1526	71	1668	193	927
928	491	1370	349	1248	652	1077	204	1657	62	1535	801
931	1218	367	210	1663	654	1086	79	1506	1375	498	787

Figure 3.58
Front face of the order-12 cube in Figure 3.45.

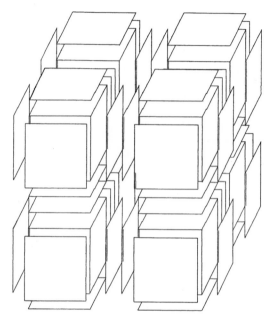

Figure 3.59
*Each pantriagonal magic cube is encased by six wrappers
consisting of six order-4 pandiagonal magic squares.*

Let's take a closer look at the various pandiagonal magic square wrappers (Fig. 3.59) in this incredible object. You can slide four of them out of the large cube (along any row, column, or pillar) and then stack them together to form a semimagic cube. For example, Figure 3.60 shows four such squares from layers 1, 6, 7, and 12 that stack to form a semimagic cube. Pretend for a moment that you are a magician who can easily shuffle the wrapper squares at will. You can interchange the layers of this cube and the result will still be a semimagic cube. Rotations and reflections of each layer are also possible so long as they all change in unison. The front face of the cube may be shifted to the back and the left face to the right in any order any time. Given that there are eight sub-cubes, this means you may view this cube in 4!(16)8 = 3072 ways.[96]

Note there are four stacks of wrapping magic squares in the vertical and eight in the horizontal similarly arranged with the same properties. Since the magic sum is the same, an entire group of pandiagonal squares can be swapped with one of the others giving 12! = 479,001,600 pos-

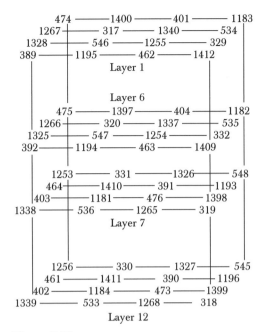

Figure 3.60
A semimagic cube formed by stacking four of the
wrapper magic squares in Figure 3.59.

sible ways of permuting these wrappers. Thus, because of the pandiagonal magic squares, you can create $12!(3072)^{12} = 3.38369 \times 10^{50}$ magic cubes. The huge number is more than the number of snow crystals formed in the ice age, which is 10^{30} crystals.

The pantriagonal subcubes allow any element to be a corner and produce forty-eight aspects due to rotations and reflection–which means you can view the cubes in 3072 ways. There are $8! = 40{,}320$ ways of interchanging the inner cubes. For these subcubes, this allows $8!(3072)^8 = 3.19808 \times 10^{32}$ different arrangements.

The entire cube, frame, and parts combined can be rotated and reflected in forty-eight aspects. This will allow an omnipotent God to make

$$48(12!)(3072)^{12}(8!)(3072)^8 = 4.32854 \times 10^{83}$$

different cubes, which exceeds the number of atoms in the universe. It even exceeds the number of electrons, protons, and neutrons in the universe, which is estimated to be 10^{79}. Despite this huge number, the odds of randomly finding a cube of this type are slim (2.5×10^{4762} to 1 against finding this particular model) because there are 1728! ways of placing 12^3 numbers in 12^3 positions.[97]

Amazing Coincidences?

Allan William Johnson, Jr., has discovered some amazing coincidences involving the two related, pandiagonal, order-4 magic squares shown here.[98]

277	277	601	431
631	401	307	197
167	337	491	541
461	571	137	367

Yin

277	197	631	431
661	401	307	167
137	337	491	571
461	601	107	367

Yang

He calls the squares Yin and Yang. Each square is composed of sixteen different three-digit prime numbers. Notice that Yin and Yang have the same main diagonals and the same magic sum 1536.

Adding 30 to each of the primes in Yin and Yang produces two more pandiagonal order-4 magic squares, Zin and Zang, each composed of sixteen different three-digit primes:

307	257	631	461
661	431	337	227
197	367	521	571
491	601	167	397

Zin

307	227	661	461
691	431	337	197
167	367	521	601
491	631	137	397

Zang

Zin and Zang are magic squares that have the same main diagonals and the same magic sum 1656. Yin, Yang, Zin, and Zang all have prime numbers with three digits. Can we find another constant to add to the squares to produce a four-digit magic square? Yes. Allan Johnson adds 1092 to each prime in Zang to yield Super-Zang with sixteen different four-digit primes arranged in a pandiagonal order-4 magic square of magic sum 6024:

1399	1319	1753	1553
1783	1523	1429	1289
1259	1459	1613	1693
1583	1723	1229	1489

Super-Zang

I do not know if it is difficult to find a constant such that when it is added to each of the numbers in Super-Zang it creates another magic square, "Ultima-Zang," with different prime numbers. So far no one has attempted such a feat.

N-Queens Solutions from Magic Squares

Ever wonder if magic squares can be used for solving practical problems? Well, they can be quite valuable in thinking about certain chess solutions. For example, consider the N-queens problem that grew from the famous 8-queens problem, which asks how does one place eight chess queens on a chessboard in such a way that no queen attacks any of the other queens. The N-queens problem is the generalization of this chessboard problem–the goal is to place N chess queens on an $N \times N$ board, subject to the same constraints. One solution of an 11-queens problem is given here:

A solution to the N-queens problem can be represented by the following permutation:[99]

1	2	...	i	...	N
a_1	a_2	...	a_i	...	a_N

Coordinates

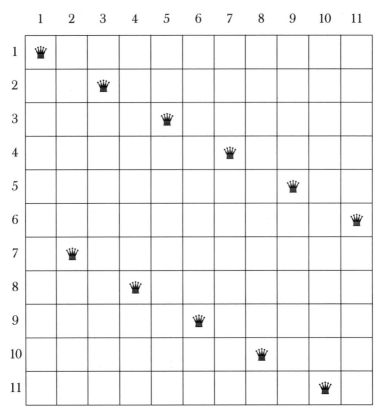

11-queens problem

in which the queens are located in coordinates (a_i, i) where i corresponds to the row number and a_i corresponds to the column number of the board. For example, the solution just illustrated for the 11-queens problem is represented as

1	2	3	4	5	6	7	8	9	10	11
1	3	5	7	9	11	2	4	6	8	10

11-queens solution

or more succinctly as (1 3 5 7 9 11 2 4 6 8 10).

It is possible to obtain solutions for the N-queens problem from magic squares constructed by de la Loubére's method as follows:[100]

1. Apply modulo N operations to every integer in the magic square X to obtain matrix Y, i.e., if x_{ij} and y_{ij} represent the

integers in the ith row and jth column of X and Y, respectively, then

$$y_{ij} = x_{ij} \bmod n$$

The mod function (or modulo function) yields the remainder after division. A number $x \bmod n$ gives the remainder when x is divided by n. This number is anywhere from 0 to $n-1$. For example, $200 \bmod 47 = 12$ because $200/47$ has 12 as a remainder.

2. Replace all the 0s in the transformed matrix Y with Ns to obtain matrix Z, i.e.,

$$z_{ij} = N \text{ if } y_{ij} = 0, \ y_{ij} \text{ otherwise}$$

Rows of the resultant matrix Z provide solutions to the N-queens problem.

Here's an example. Let's construct an 11×11 magic square using de la Loubére's method:

68	81	94	107	120	1	14	27	40	53	66
80	93	106	119	11	13	26	39	52	65	67
92	105	118	10	12	25	38	51	64	77	79
104	117	9	22	24	37	50	63	76	78	91
116	8	21	23	36	49	62	75	88	90	103
7	20	33	35	48	61	74	87	89	102	115
19	32	34	47	60	73	86	99	101	114	6
31	44	46	59	72	85	98	100	113	5	18
43	45	58	71	84	97	110	112	4	17	30
55	57	70	83	96	109	111	3	16	29	42
56	69	72	95	108	121	2	15	28	41	54

Eleventh-order magic square

Applying the mod 11 operation to this magic square, we get the resulting square:

2	4	6	8	10	1	3	5	7	9	11
3	5	7	9	11	2	4	6	8	10	1
4	6	8	10	1	3	5	7	9	11	2
5	7	9	11	2	4	6	8	10	1	3
6	8	10	1	3	5	7	9	11	2	4
7	9	11	2	4	6	8	10	1	3	5
8	10	1	3	5	7	9	11	2	4	6
9	11	2	4	6	8	10	1	3	5	7
10	1	3	5	7	9	11	2	4	6	8
11	2	4	6	8	10	1	3	5	7	9
1	3	5	7	9	11	2	4	6	8	10

Eleventh-order magic square mod 11

Reading each row from left to right and right to left in this figure provides 11-queens permutations. Note that the last row corresponds to the solution shown in the previous 11 × 11 board with queens on it. We can obtain other solutions by considering the principal major diagonal (top left to bottom right), that is, (2, 5, 8, 11, 3, 6, 9, 1, 4, 7, and 10), and other major diagonals, such as (8, 11, 3, 6, 9, 1, 4, 7, 10, 2, and 5), which is highlighted in this mod-11 square.

The square arrangements of *N*-sets of nonattacking queens on an *N* × *N* board with no overlaps is sometimes called Latin queens squares. You can also see that if we were to place queens at all the 1s in the previous board, they would all be nonattacking. The same is true for the set of 2s, the set of 3s, and so forth. Here is a board showing queens at 1 and 7 cells.

2	4	6	8	10	♛	3	5	♛	9	11
3	5	♛	9	11	2	4	6	8	10	♛
4	6	8	10	♛	3	5	♛	9	11	2
5	♛	9	11	2	4	6	8	10	♛	3
6	8	10	♛	3	5	♛	9	11	2	4
♛	9	11	2	4	6	8	10	♛	3	5
8	10	♛	3	5	♛	9	11	2	4	6
9	11	2	4	6	8	10	♛	3	5	♛
10	♛	3	5	♛	9	11	2	4	6	8
11	2	4	6	8	10	♛	3	5	♛	9
♛	3	5	♛	9	11	2	4	6	8	10

Nonattacking queens

I've just shown how it is possible to obtain *N*-queens solutions from magic squares. Researchers have also used *N*-queens solutions to perform the reverse operation–constructing magic squares–and this is described in the literature that also discusses a modification of the *N*-queens problem called the *N*-super-queens problem in which you must place *N* queens on a *toroidal* board.[101]

Palindromes and Magic Squares

The following is an example of sixteen different palindromes that can be arranged in a 4×4 magic square whose magic sum is also a palindrome.[102] (A palindrome is a number that reads the same backward and forward.) This square with magic sum of 393 was discovered in 1989 by brute-force computation using a computer.

2	252	131	8
99	101	171	22
111	33	88	161
181	7	3	202

Palindrome magic square

Here is another palindromic magic square that has the added property of having main diagonals containing eight repeating "repdigits": 222, 333, 444, 555, 666, 777, 888, and 999. The magic sum is 2442.

222	595	737	888
959	666	444	373
484	333	999	626
777	848	262	555

Diagonal repdigits

Can you find larger palindromic magic squares?

The Swastika Magic Square

The swastika magic square is such that the rows, columns and two diagonals add up to 65, and all the prime numbers that occur between 1 and 25 (1, 2, 3, 5, 7, 11, 13, 17, 19, and 23) can be found within the swastika shape.[103]

17	5	13	21	9
4	12	25	8	16
11	24	7	20	3
10	18	1	14	22
23	6	19	2	15

Swastika magic square

Do there exist other swastika magic squares? Can you find swastikas for squares of higher order?

The 1234 Square

The following is a rather odd magic square. Every cell contains the same number, 1234. I know it must look fairly boring to you.

1234	1234	1234
1234	1234	1234
1234	1234	1234

1234 Square

Although rather benign looking, "1234" has some rather horrific connotations. In particular, 1234 was the year Pope Gregory IX established the dreaded Inquisition. Centuries later, your challenge is to place nine different four-digit numbers (using the same digits) that also form a magic square.[104] Altogether, there must be nine 1s, 2s, 3s, and 4s. Here is one beautiful solution:

2243	1341	3142
3141	2242	1343
1342	3143	2241

1234 solution

How many other solutions can you find? Do any solutions exist that retain the year the Inquisition was established?

Hydrogen-Bond Codon-Anticodon Magic Square

DNA contains the basic hereditary information of living cells and is expressed as a four-letter code symbolized by the letters G, C, A, and T. These letters stand for the chemical bases guanine, cytosine, adenine, and thymine. The genetic message is composed of a string of these bases and is transcribed to form a related molecule, RNA, whose base sequence is complementary to the DNA. (In RNA the letter U stands for uracil, which takes the place of T in DNA.) The chemical bases in DNA hydrogen-bond to the complementary bases in RNA. Cs bond to Gs and As bond to Us. For instance, the codon triplet GCG binds to its complementary anticodon CGC using nine hydrogen bonds–three bonds between each complementary pair. Generally speaking, the more bonds, the more tightly coupled are the codon and anticodon. The set of sixty-four codons is sometimes referred to as life's genetic code.

Amino acids, the basic building blocks of proteins, are coded for by specific consecutive triplets of these bases. The triplets are called codons. For example, GGG codes for the amino acid glycine.

Researcher Gary Adamson from San Diego, California, has discovered that the set of three-letter codons in the genetic code corresponds to a magic square, and the number of hydrogen bonds for each codon-anticodon pair can be computed by assigning two numbers to each base as follows:[105]

Base:	C	G	A	T/U
Two-digit	0	1	0	1
code:	0	1	1	0

For example, we can convert the codon CUG to

C	U	G
0	1	1
0	0	1

In order to form the final magic square, we first have to convert a codon triplet to a single number by determining the number of bit changes between the first and second row of bits. For example, for CUG we find there is a 1-bit change between the top row (011) and the bottom row (001). Information theorists sometimes refer to the number of bit changes as the *Hamming distance.* All Hamming distances (in this case a 1) are subtracted from 9. This means we get a value of 8 corresponding to CUG, and 8 happens to be the number of hydrogen bonds for the CUG-GAC pair. Using this approach we can create what Adamson calls an *H-bond codon-anticodon magic square* with magic constant 60. This square lists all of nature's sixty-four possible codons next to their numerical code produced by the aforementioned recipe.

CCC 9	CCU 8	CUU 7	CUC 8	UUC 7	UUU 6	UCU 7	UCC 8
CCA 8	CCG 9	CUG 8	CUA 7	UUA 6	UUG 7	UCG 8	UCA 7
CAA 7	CAG 8	CGG 9	CGA 8	UGA 7	UGG 8	UAG 7	UAA 6
CAC 8	CAU 7	CGU 8	CGC 9	UGC 8	UGU 7	UAU 6	UAC 7
AAC 7	AAU 6	AGU 7	AGC 8	GGC 9	GGU 8	GAU 7	GAC 8
AAA 6	AAG 7	AGG 8	AGA 7	GGA 8	GGG 9	GAG 8	GAA 7
ACA 7	ACG 8	AUG 7	AUA 6	GUA 7	GUG 8	GCG 9	GCA 8
ACC 8	ACU 7	AUU 6	AUC 7	GUC 8	GUU 7	GCU 8	GCC 9

Hydrogen-bond codon-anticodon magic square

In actuality, the square is only semimagic because the left diagonal is all 9s and the right diagonal is all 8s.

Is it remarkable that conversion of the genetic code to a number of bit changes conveniently represents the number of hydrogen bonds? You may be asking yourself how the table was arranged to form the magic square in the first place. A recipe is on page 289.

	000	**001**	**011**	**010**	**110**	**111**	**101**	**100**
000	000 000 CCC 9	001 000 CCU 8	011 000 CUU 7	010 000 CUC 8	110 000 UUC 7	111 000 UUU 6	101 000 UCU 7	100 000 UCC 8
001	000 001 CCA 8	001 001 CCG 9	011 001 CUG 8	010 001 CUA 7	110 001 UUA 6	111 001 UUG 7	101 001 UCG 8	100 001 UCA 7
011	000 011 CAA 7	001 011 CAG 8	011 011 CGG 9	010 011 CGA 8	110 011 UGA 7	111 011 UGG 8	101 011 UAG 7	100 011 UAA 6
010	000 010 CAC 8	001 010 CAU 7	011 010 CGU 8	010 010 CGC 9	110 010 UGC 8	111 010 UGU 7	101 010 UAU 6	100 010 UAC 7
110	000 110 AAC 7	001 110 AAU 6	011 110 AGU 7	010 110 AGC 8	110 110 GGC 9	111 110 GGU 8	101 110 GAU 7	100 110 GAC 8
111	000 111 AAA 6	001 111 AAG 7	011 111 AGG 8	010 111 AGA 7	110 111 GGA 8	111 111 GGG 9	101 111 GAG 8	100 111 GAA 7
101	000 101 ACA 7	001 101 ACG 8	011 101 AUG 7	010 101 AUA 6	110 101 GUA 7	111 101 GUG 8	101 101 GCG 9	100 101 GCA 8
100	000 100 ACC 8	001 100 ACU 7	011 100 AUU 6	010 100 AUC 7	110 100 GUC 8	111 100 GUU 7	101 100 GCU 8	100 100 GCC 9

Codons mapped to gray code

Write down a row of eight *binary gray-code* numbers (binary gray-code numbers have a single digit differing by 1). We can compute the nth gray code term from the $(n-1)$th term in a simple manner. If the $(n-1)$th term has an even number of 1s, change the right-most bit in n; if the $(n-1)$th term has an odd number of 1s, change the bit to the left of the rightmost digit to 1. This recipe produces a "gray code" sequence: 000, 001, 011, 010, 110, 111, 101, 100.... Next, write out the eight gray codes as rows and columns. At the corresponding array cell positions, write the row bits beneath the column bits. To the right of each pair, write out the modified Hamming distance (number of bit changes) between the pairs. For example, (0, 1, 1) and (0, 0, 0) have two bit changes, which we then subtract from 9 to get the value 7 in the table. In this manner, we create a semimagic square that is a collection of these values.

The table on page 288 has many hidden symmetries waiting for you to discover. For example, you can draw a mirror plane through the left diagonal, and all hydrogen-bond numbers will be identical on both sides of the mirror plane. There is also a one-unit difference between adjacent cells (up, down, right, and left, including wrap-arounds) and a one-letter difference between adjacent codons.

Mystery Pattern

Figure 3.61 shows a mysterious pattern that I give readers to ponder. Can anyone determine the origin of the pattern and the significance of the symbols?[106] (A solution is given in note 106.)

Magic Dice and Dominoes

In 1999, another unusual magic configuration was invented by G. L. Honaker of Bristol, Virginia, as a math enrichment puzzle for his high-school students.[107] The object of Honaker's original puzzle was to number the pips (dots) on an ordinary six-sided die with distinct positive integers such that the numbers on each face add up to the same magic constant. What is the smallest possible magic sum?

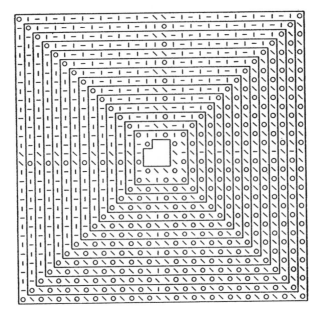

Figure 3.61
Mystery pattern. Can you determine the relevance of the symbols?

One solution that achieves the minimum sum of 43 is shown in Figure 3.62. This arrangement of numbers can also be represented as a numerical "magic triangle."

                        43

                   21   22

               8   16   19

           5   9   12   17

       3   4   7   14   15

   1   2   6   10   11   13

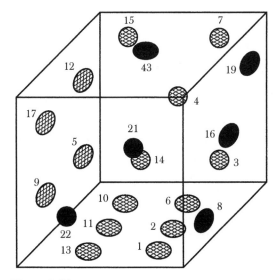

Figure 3.62
A magic numbering of an ordinary die with magic sum 43.

The sum of each row is required to be the magic constant. The number at the apex of the triangle is the number on the face with one pip. The two numbers in the next row are the two numbers on the face with two pips (listed in any order) and so on for the other rows.

Here is a simple proof that 43 is the smallest possible magic sum. Consider the bottom four rows of the triangle, indicated by the o symbols:

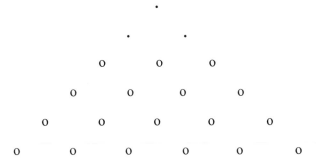

There are 18 locations to be filled, so the smallest possible sum is $1 + 2 + \cdots + 18 = 171$, which occurs if we fill the locations with the numbers 1 to 18 in some order. Since each of these four rows must have the same sum M, we have $4 \times M = 171$; hence $M = 42.75$. Because the magic sum must be an integer, this argument shows the magic sum is at least 43. Since there is a solution with $M = 43$, this proves that 43 is minimal.

This problem can be generalized to numbering an n-sided die. Mathematicians Mike Keith and Judson McCranie have shown that the magic sum in this case is at least

$$M(n) = \text{ceiling}(T(T(n) - T(\lfloor n/3 \rfloor)) / (n - \lfloor n/3 \rfloor))$$

where $T(n)$ is the nth triangular number, $n(n + 1) / 2$; $\lfloor x \rfloor$ means to round a number down to the nearest integer, and "ceiling(x)" means to round a number up to the nearest integer. Keith and McCranie have successfully constructed solutions with exactly $M(n)$ as the magic sum for all n up to several hundred, and thus they conjecture that $M(n)$ is the minimal magic sum for all n. A proof of this conjecture is not yet known. In addition to these cubical dice, you might enjoy attempting to construct optimal magic numberings for dice made from the other four platonic solids (with 4, 8, 12, and 20 sides), whose magic sums are 15, 94, 294, and 1283.

Several other variations are possible. More than one die can be simultaneously numbered with distinct integers such that each die is magic but each die has a different magic sum. In this case the goal is to minimize the largest magic sum of any die. For two six-sided dice the minimum possible sum is 84, as illustrated by

		84			
	41		43		
	22	24	38		
7	11	32	34		
2	8	16	28	30	
4	9	10	15	20	
		26			

		83			
	39		44		
	23	25	35		
5	14	31	33		
3	6	18	27	29	
1	12	13	17	19	2
		1			

Other arrangements of pips can be considered in which all the pips on the twenty-seven dominoes of a double-six set (minus the double-blank) have been numbered such that the sum on each domino is the same and no numbers repeat. McCranie, Keith, and Honaker have recently proven that the magic sum of 564 is the smallest possible.

In the following table, the domino with a single pip (not shown) would be labeled 564. The set of numbers used to label each pip cannot be consecutive. Note that other numberings with this same sum are possible. In these kinds of magic dice and dominoes, it is also possible to find magic numberings that yield very large magic sums.

Domino	$S = 564$	Domino	$S = 564$	Domino	$S = 564$
	281 + 283		187 + 188 / 189		135 + 136 / 145 + 147
	137 + 138 + 144 / 145		101 + 103 + 119 / 120 + 121		104 + 109 + 116 + 117 / 118
	78 + 79 + 81 / 107 + 108 + 111		82 + 84 + 85 + 102 / 105 + 106		88 + 89 + 99 + 90 + 98 / 100
	47 + 56 + 48 / 150 + 131 + 77 + 55		83 + 86 + 71 + 87 + 94 / 75 + 68		162 + 156 + 126 + 70 + 30 + 15 / 5
	133 + 155 + 76 + 59 / 43 + 37 + 35 + 26		153 + 148 + 52 + 123 + 65 / 14 + 7 + 2		72 + 73 + 74 + 80 + 63 + 66 / 67 + 69
	154 + 141 + 18 + 130 + 40/22 + 11 + 32 + 16		152 + 128 + 61 + 58 + 45 +33 / 31 + 29 + 27		134 + 132 + 8 + 129 + 4 / 12 + 3 + 62 + 39 + 41
	151 + 143 + 50 + 1 + 10 + 24 / 34 + 60 + 49 + 36		127 + 125 + 54 + 44 + 28 + 21 / 6 + 3 + 46 + 51 + 53		13 + 17 + 19 + 20 + 23 + 25 / 38 + 42 + 57 + 64 + 122 + 124

When labeling the dominoes, each pip must have a number. For example, in the upper left of the table, the two pips are assigned 291 and 283, which give a sum of 564. A "/" symbol is used to delineate one square of the domino from the other. Not all dominoes are shown in the table. Here is a challenging puzzle: can you assign values to the missing dominoes?

In 1999, another kind of magic die was invented by Jeremiah Farrell of Indianapolis, Indiana.[108] It is possible to arrange twenty-seven dice in a $3 \times 3 \times 3$ cube so that the faces have special properties. On the next page is what the cube looks like when unfolded to

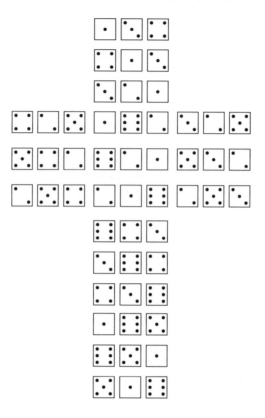

reveal all of the face values. Notice that the overall magic die is magic on every face with respect to the rows and columns. In addition, when the magic die is placed on the table, the die is magic around the four sides (ignore the top and bottom) in a special way. Here's how it works. Create the magic die by folding up the T-shaped diagram. Choose one of the diagonals, rows, or columns that you see on the side of the die. Then add the four corresponding diagonals, rows, or columns around the four faces, and the total will always be 42. For example, we might choose the face with 4-3-1 on the top row. Next, we add the corresponding top row around the four sides: $4 + 3 + 1 = 8$ plus $2 + 1 + 6 = 9$, plus $3 + 4 + 6 = 13$, plus $5 + 1 + 6 = 12$. The total is 42.

The properties of this magic die become easier to understand if you photocopy the diagram, paste it on heavy paper, and then fold it into a cube. You can use transparent tape to hold it together. You can also construct a magic die puzzle by taking twenty-seven dice and gluing them together. There will be only one magic solution, the solution with magic constant 42.

Here is another example. First construct the cube, and then place the cube on a table so that the face with the 3-3-3 diagonal is on top and the face with the 4-4-4 diagonal is on the bottom. Next, we can choose to examine a row, column, or diagonal. Once the choice is made, we add together the corresponding choices around the four sides. For example, choose row $3 + 4 + 6 = 13$. In fact, position this die face so that it faces toward you. Rotate the cube to show the corresponding face with $5 + 1 + 6 = 12$. Rotate the cube so that the face with $4 + 3 + 1 = 8$ is forward. Rotate again so that $2 + 1 + 6 = 9$ is forward. The four sums sum to 42.

Here is another example. We can use the main diagonals, $4 + 1 + 3 = 8, 2 + 2 + 2 = 6, 3 + 6 + 4 = 13$, and $5 + 5 + 5 = 15$. Again, the four sums sum to 42. The corresponding diagonals are hard to visualize on the unfolded T-diagram but easy to find if you construct the magic die so that you can place it on a table and turn it.

Jeremiah Farrell is emeritus professor of mathematics at Butler University in Indianapolis but still teaches an occasional class on special topics such as combinatorial games. He now spends his time inventing mathematical and linguistic games and puzzles. In 2000 he mailed me a marvelous physical model of his magic die made of twenty-seven dice glued together. Today it sits proudly on my desk next to an assortment of other magical figures.

CHAPTER FOUR

Gallery 2: Circles and Spheres

*Laws of physics and mathematics are like a coordinate system that
runs in only one dimension. Perhaps there is another dimension
perpendicular to it, invisible to those laws of physics, describing the
same things with different rules, and those rules are written in our
hearts, in a deep place where we cannot go and read them except in
our dreams.*

–Neal Stephenson, *The Diamond Age*

*Mathematical inquiry lifts the human mind into closer proximity
with the divine than is attainable through any other medium.*

–Hermann Weyl

*We sail within a vast sphere, ever drifting in uncertainty, driven from
end to end.*

–Blaise Pascal

This chapter contains an exhibition gallery of my favorite magic
objects involving circles and spheres. Although circles and spheres
have been discovered with incredible beauty and complexity, these
kinds of objects have been studied much less than magic squares
and cubes. Perhaps one reason for the relative paucity of research is
that the circles and spheres are more difficult to represent on paper
than the squares and cubes consisting of tables of numbers. Perhaps
the twenty-first century, with its increasing use of virtual-reality computer tools and 3-D graphics, will be a renaissance in our understanding of magic circles, spheres, and even four-dimensional
hyperspheres. I look forward to the adventure. If many of us could
develop the ability to mentally visualize magic hyperspheres with
startling properties, could this alter our worldview and stimulate

additional interest in magic objects? The history of mathematics is replete with the acceptance of concepts beyond our imagination. But that doesn't mean it will be impossible to visualize higher-dimensional structures. As Edward Kasner and James Newman note in *Mathematics and the Imagination,* "For primitive man to imagine the wheel, or a pane of glass, must have required even higher powers than for us to conceive of a fourth dimension." During our European Renaissance, new knowledge flooded medieval Europe with the light of intellectual transformation, wonder, creativity, exploration, and experimentation. Another mini-renaissance in recreational mathematics could be fueled by a wealth of higher-dimensional magic spheres and stars. Interestingly, the spirit of our European Renaissance achieved its sharpest formulation in art. Art came to be seen as a branch of knowledge, valuable in its own right and capable of providing both spiritual and scientific windows to our vast universe. Similarly, the excitement caused by high-dimensional magic spheres and stars might transform art and mathematics with new ideas, forms, and emotions.

Before exploring the exciting arena of magic spheres, let's start with simpler constructs on a plane. For example, groups of circles may be arranged so that their intersection points have magical properties.[1] In most of the figures that follow, the sum of the integers on any circle is equal to the sum of the integers on any other circle.

Magic Circular Die and Squashed Die

Next time you are near a die, take a look at its six sides. You'll notice the 6 is opposite the 1, 5 is opposite 2, and 4 is opposite 3 – each complementary pair summing to 7. This means that any band of four numbers encircling the die sum to 14.

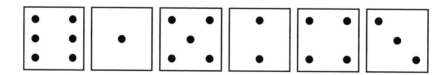

The word "quincunx" is the name for the pattern

on a die. It's also the name for a particular type of five-domed cathedral, such as St. Mark's Cathedral in Venice. (Certain Khmer temples in southeast Asia also use this configuration.)

Figure 4.1*a* is a spherical representation of a die, and Figure 4.1*b* is what happens when you sit on the die and squash it like a bug.[2]

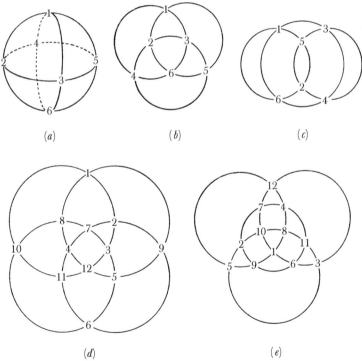

Figure 4.1
Magic circles.

This is the simplest form of a magic circle in which the numbers along each circle sum to a constant 14. Figure 4.1*c* is another construction giving the same results as Figure 4.1*b*. Any pair of complementary numbers is common to two circles. For example, in Figure 4.1*b*, 6 and 1 both reside on the left circle and the right circle.

The circle assembly in Figure 4.1*d* contains the numbers 1 through 12 arranged in four circles. Each circle has six numbers with magic sum of 39. Figure 4.1*e* is another magic assembly of four circles.

Figure 4.2 shows magic circles for assemblies of five and six circles.

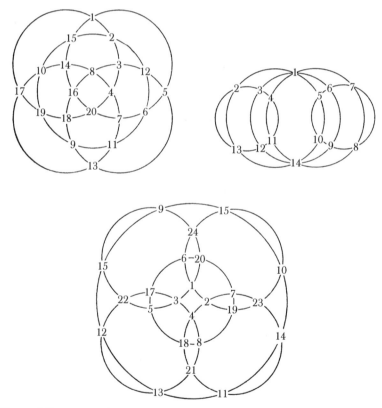

Figure 4.2
Magic circles with five and six rings.

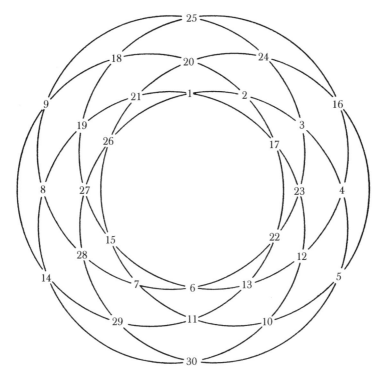

Figure 4.3
Magic six-circle with special diameters.

Magic Six-Circle

Figure 4.3 shows six circles of ten numbers, each circle with magic sum 155. The numbers 1 through 30 are used.[3] Also, if you add together the four and six numbers on any two diameter lines, you get the magic sum of 155. For example, $(25 + 20 + 1 + 6 + 11 + 30) + (18 + 21 + 13 + 10) = 155$.

Magic Eight-Circle

Figure 4.4 shows eight circles arranged so that they each have ten intersections.[4] The magic sum is 205 for each circle. The consecutive numbers 1 through 40 are used. Beautiful!

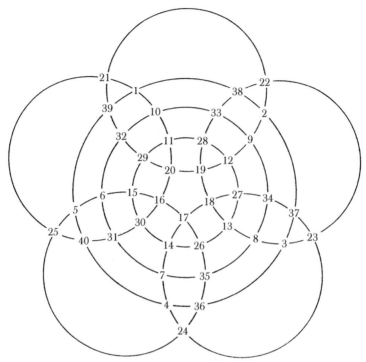

Figure 4.4
Magic eight-circle.

Uncommon Magic Circle

Figure 4.5 contains the numbers 1 through 8 arranged in eight circles of four numbers each with magic constant 18.[5] There's some-

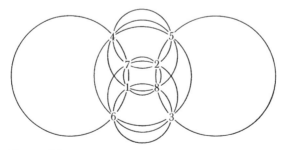

Figure 4.5
Magic eight-circle. Each number is at the intersection of four circles but no other point is common to the same four circles.

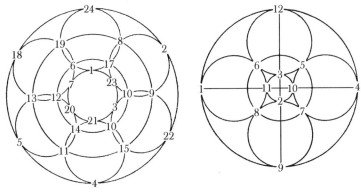

Figure 4.6
Magic ten-circle.

Figure 4.7
Magic seven-circle.

thing strange about this set of circles that sets it apart from the previous ones. Can you tell? It turns out that each number is at the intersection of four circles but no other point is common to the same four circles.

Magic Seven-Circle

Figure 4.6 contains the numbers 1 though 24 arranged in ten circles of six numbers each with magic constant 75.[6] Figure 4.7 contains the numbers 1 though 12 arranged in seven circles and two diameter lines, of four numbers each, with magic constant 26.

A Glorious Magic Set of Nine Circles

Figure 4.8 contains the consecutive numbers 1 through 54 arranged in nine circles.[7] Each circle contains twelve numbers with a magic sum of 330. And here is the icing on the cake: The six 3×3 number clusters form magic squares!

A Glorious Magic Set of Twelve Circles

Figure 4.9 contains the consecutive numbers 1 through 96 arranged in twelve circles.[8] Each circle contains sixteen numbers

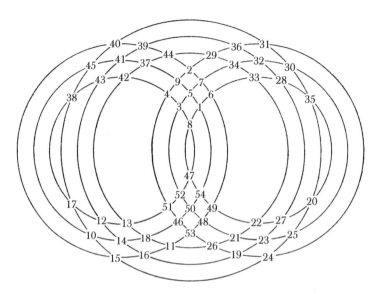

Figure 4.8
Magic nine-circles with six inlaid magic squares.

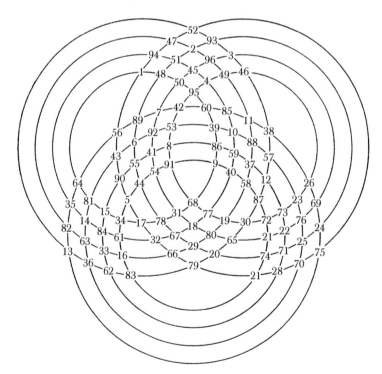

Figure 4.9
Magic twelve-circles with inlaid fourth-order pandiagonal magic squares.

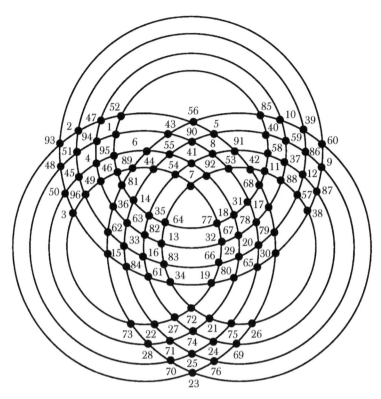

Figure 4.10
Magic twelve-circles with inlaid fourth-order pandiagonal magic squares.

with the magic sum of 776, the year that Muslim scientist Al-Jahiz[9] was born. Better yet, the numbers in each of the six 4×4 clusters of numbers sum to 776, and each is a Jaina square constructed by Phillipe de la Hire's method as described in the Introduction.

Figure 4.10 is a modification of these circles made by Harvey Heinz.[10] We see six 4×4 pandiagonal magic squares that each sum to 194 in fifty-two different ways (four rows, four columns, two main diagonals, six broken diagonal pairs, corners of four 3×3 squares, corners of sixteen 4×4 squares including wrap-around, and sixteen 2×2 squares, including wrap-around). The twelve circles of sixteen numbers each sum to 776 (4×194).

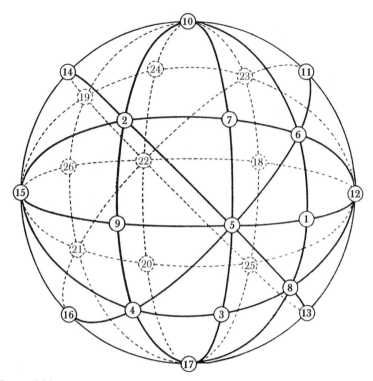

Figure 4.11
Nine-circle magic sphere.

Nine-Circle Magic Sphere

Figure 4.11 is a sphere containing the consecutive numbers 1 through 26 arranged in nine circles.[11] Each circle has eight numbers, and their magic sum is 108. Notice that pairs of numbers on each side of an imaginary diameter line sum to 27.

Seven-Circle Magic Sphere

Figure 4.12 is a sphere containing the consecutive numbers 1 through 26 arranged in seven circles.[12] Each circle has eight numbers, and their magic sum is 108.

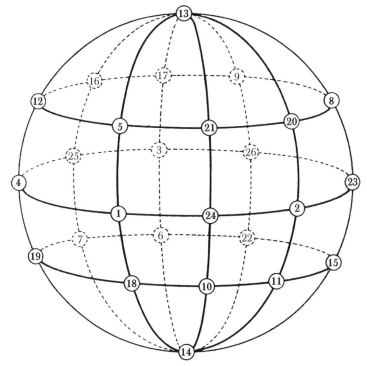

Figure 4.12
Seven-circle magic sphere.

Eleven-Circle Magic Sphere

Figure 4.13 is a sphere containing the consecutive numbers 1 through 62 arranged in eleven circles.[13] Each circle has twelve numbers, and their magic sum is 378.

Concentric Onion Spheres

Figure 4.14 shows two concentric spheres with the consecutive numbers 1 through 12 arranged in six circles.[14] Each circle contains four numbers with magic sum 26. There are also three diameter lines running through the spheres with magic sum 26.

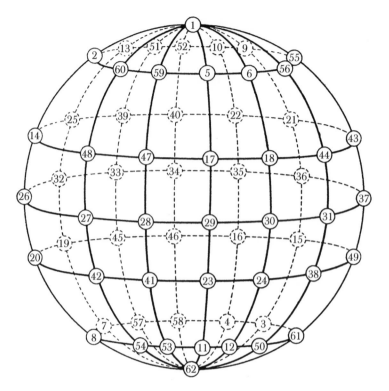

Figure 4.13
Eleven-circle magic sphere.

Figure 4.15 is similar to the previous concentric spheres except that, sadly, two of the circles do not give the magic constant 26.[15] But this small sacrifice allows us to create twelve additional summations of 26 that are shown by the dotted circles in Figures 4.16–4.18. Figure 4.16 shows the vertical receding plane of eight numbers. Figure 4.17 shows the horizontal plane, and Figure 4.18 shows the plane parallel to the paper, the latter continuing the two concentric circles that do not sum to 26. Figure 4.19 shows a geometrical diagram. The solid lines show the selections for Figure 4.16, and the dotted lines for Figure 4.17.

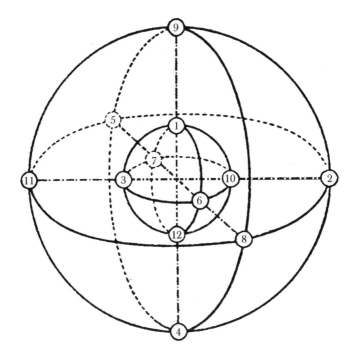

Figure 4.14
Onions, or spheres within spheres.

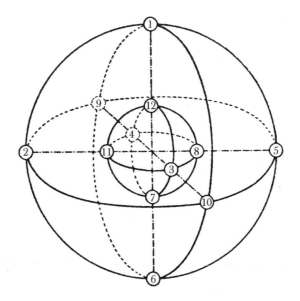

Figure 4.15
Onions, or spheres within spheres.

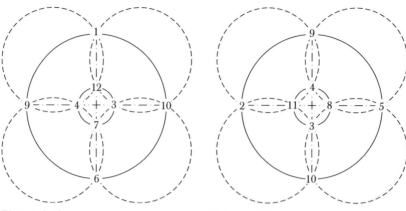

Figure 4.16
Vertical receding plane for Figure 4.15.

Figure 4.17
Horizontal plane.

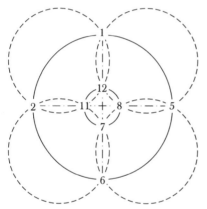

Figure 4.18
Plane parallel to paper.

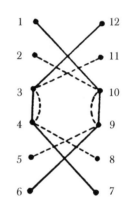

Figure 4.19
Geometrical diagram.

Fifteen-Circle Magic Sphere

Along with John Hendricks's ornate, inlaid magic cubes in chapter 3, Figure 4.20 is my favorite magic object in this book. It's a sphere containing the consecutive numbers 1 through 98 arranged in fifteen circles.[16] Each circle has sixteen numbers, and their magic sum is 792. Moreover, there are six 3×3 magic squares. Two of these

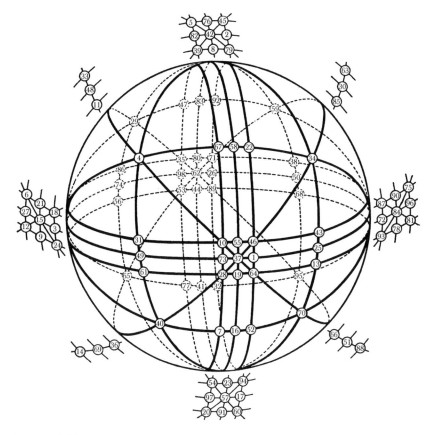

Figure 4.20
Fifteen-circle magic sphere with inlaid third-order magic squares.

magic squares can form the nucleus of a 5×5 concentric square. Also, the sum of any two diametrically opposite numbers is 99.

Magic Circle-Square Hybrid

The two diagrams in Figure 4.21 illustrate some relationships in the following order-4 magic square.[17] The rows, columns, diagonals, and circular connections between the numbers sum to 34. The two diagrams represent the same square. Notice how useful it is to indicate relationships with circles so that hidden patterns become obvious to the eye.

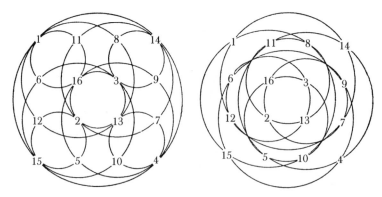

Figure 4.21
Magic circle-square hybrid.

1	11	8	14
6	16	3	9
12	2	13	7
15	5	10	4

Magic circle-square hybrid

Samsara Circles

I presented the following interlocking *Samsara circles* on the Internet as a tough challenge for colleagues to solve (Fig. 4.22).[18] As part of my Grand Internet Math Challenge, friends were asked if it were possible to place consecutive numbers from 1 to 10 at the circles' intersection points so that the numbers along each circle would sum to a constant. Congratulations to Michael Keith, a software engineer from Salem, Oregon, the first person on Earth to find a partial solution to the mysterious Samsara circles. He received a free autographed copy of my novel *Spider Legs* for his effort, which is summarized below.

Take a look at Figure 4.22. At first glance, it does not appear possible to use consecutive integers to solve this problem. There-

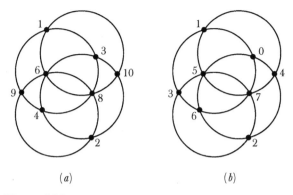

Figure 4.22
Samsara circles. (a) Keith solution. (b) Anderson solution.

fore, we will discuss the solution using the smallest set of positive integers. The four-circle set has eight intersection points, so the best one could hope for is a solution using {1–8}. Sadly, such a solution does not exist, nor is there a solution using numbers taken from {1–9}, but there are quite a few solutions if we use numbers 1 through 10. In Mike's solution (Fig. 4.22a), the sum for each circle is 25.

Do you think the four-circle problem could be solved with consecutive numbers in which the initial number is greater than 1? We believe it is impossible. The key to understanding this is to note that two of the circles use five intersection points and two use four points. Now suppose the Samsara circles can be solved with the set {$M, \ldots, M+7$}. Imagine M is "large." Then, by the magic condition, one of the five-point circles must equal one of the four-point ones, for example, $b + c + f + g + h = b + d + e + h$ (which is, in fact, one of the equations to solve if the intersection points are labeled a through h). Next, we write $a = M + a'$, $b = M + b'$, etc., where a', b', etc. are integers in the range 0 to 7. The equation we must solve becomes $5M + (b' + c' + f' + g' + h') = 4M + (b' + d' + e' + h')$, which can be simplified to $M = (d' + e') - (c' + f' + g')$. However, we know that a', b', etc. are small numbers, and so there is no way the expression on the right-hand side can be equal to the large number M. In fact, M only needs to be 11 or larger for the equation to have no solutions because the largest the right-hand side can be is $(7 + 6) - (2 + 1 + 0) = 10$. This suggests that the Samsara circles can't be created with large values of M. The overall proof is completed by showing, using a computer, that it is also impossible to create a set of intersection points for small M, up to a limit where the previous argument kicks in. Michael Keith has tested this using a computer, so we believe that there are no consecutive positive number solutions to the Samsara circles.

If we permit the use of zero or negative numbers, there are exactly forty solutions to the Samsara rings using consecutive integers. Arlin Anderson from the U.S. Army Aviation and Missile Command at the Redstone Arsenal in Alabama notes that ten of these solutions start with zero, and thirty begin with negative numbers. Twenty of the forty solutions have positive totals and twenty have negative. Figure 4.22*b* shows an example. Can anyone find a solution such that the numbers, when multiplied together, form a magic product?

Almost the Olympics

Near-Olympian ring configurations are ripe for magic. Figure 4.23 is an original set of rings in which the magic sum is 44. Michael Keith points out that perfect solutions with consecutive integers from 1 to 14 are not possible.[19] The solution shown skips the number 10 and ends with 15.

Notice that this ring configuration is not quite the same as the Olympics logo because the rings in Figure 4.23 are more loosely linked, and cannot be magically numbered. Can you use numbers that are all prime to form a more awesome set of rings?

The Rings of Callicrates

Figure 4.24 is a wonderful original design by Michael Keith[20] and named after the Athenian architect Callicrates for its simplicity and

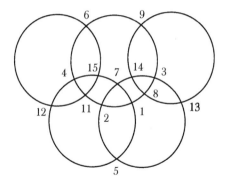

Figure 4.23
Near-Olympian rings.

beauty. The configuration is perfectly magic using the consecutive numbers 1 through 80 on eighty intersection points, and each circle adds to 324. To make this the most magical of magic rings, notice that the numbers on the outer intersection points are the digits of pi: 3.14159265358. . . . We believe that this kind of arrangement can be generalized to *N* circles in each ring.

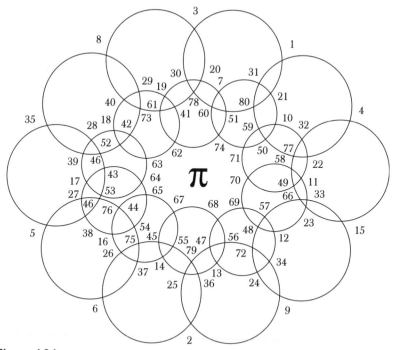

Figure 4.24
Rings of Callicrates. To make this the most magical of magic rings, notice that the numbers on the outer intersection points are the digits of pi: 3.14159265358 . . . .

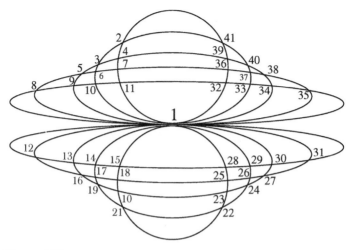

Figure 4.25
Rising sun reflected in water.

Rising Sun Reflected in Water

One beautiful example of magic geometries is the *rising sun patterns* of Michael Keith.[21] This class of patterns was researched and invented in early 1999. In Figure 4.25, each circle and ellipse has a magic sum of 173. All circles and ellipses intersect at the central 1 where the "sun" touches the water, symbolizing the "unity of nature, numbers, and the cosmos."

At first, it may seem difficult to assign numbers to this devastatingly elegant figure. However, if you look closely, you will soon see it is possible to create larger rising sun diagrams with an even number of ellipses. (Is it possible to create these figures with an odd number of ellipses?)

Small Gumdrop Configurations

Michael Keith has catalogued the following circular configurations with ⊗ intersection points.[22] All of the patterns in Figures 4.26 and 4.27 use consecutive numbers starting at 1. Can you find other solutions? We affectionately call the shapes *gumdrop configurations* because of their candylike appearance when rendered as colored circles.

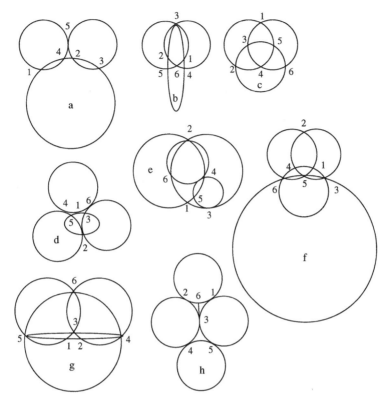

Figure 4.26

Gumdrop configurations. (a) Three circles, $\otimes$ = 5, $\mathcal{S}$ = 10. (b) Three circles, $\otimes$ = 6, $\mathcal{S}$ = 15. (c) Three circles, $\otimes$ = 6, $\mathcal{S}$ = 14. (d) Four circles, $\otimes$ = 6, $\mathcal{S}$ = 11. (e) Four circles, $\otimes$ = 6, $\mathcal{S}$ = 12. (f) Four circles, $\otimes$ = 6, $\mathcal{S}$ = 14. (g) Four circles, $\otimes$ = 6, $\mathcal{S}$ = 15. (h) Four circles, $\otimes$ = 6, $\mathcal{S}$ = 9. (Note the small line segment.)

In more technical language, Figures 4.26 and 4.27 enumerate all small magic configurations. In particular, we seek a collection of subsets of the integer set $I = \{1, 2, 3, \ldots, N\}$ with the following properties: (a) there are at least two subsets, (b) each subset has at least two elements, (c) each element of I appears in at least two subsets, and (d) elements in each subset have the same sum. By subsets we mean a collection of numbers that may be grouped in a circle or ellipse. The other conditions mean we want at least two circles each with at least two numbers.

Figure 4.26a is the smallest magic configuration and the only one with five intersection points ($\otimes$ = 5). The magic sum $\mathcal{S}$ is 10.

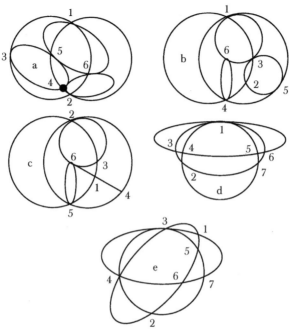

Figure 4.27
*Gumdrop configurations. (a) Five circles, ⊗ = 6, C = 12. (b) Five
circles, ⊗ = 6, C = 10. (c) Five circles, ⊗ = 6, C = 11. (d) Three
circles, ⊗ = 7, C = 19. (e) Three circles, ⊗ = 7, C = 21.*

There are ten configurations with ⊗ = 6, and these are shown in
Figures 4.26 and 4.27. Figure 4.26c is the smallest one with six-
fold symmetry. Figures 4.26d and 4.26e are the only two gumdrop
configurations with ⊗ = 7 and three circles. However, altogether
there are 271 configurations for ⊗ = 7 and various numbers of
circles:

Number of "Circles" (for ⊗ = 7)	*Number of Gumdrop Configurations*
4	23
5	98
6	105
7	38
8	5

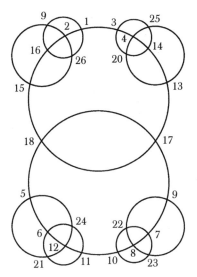

Figure 4.28
Circles of Prometheus.

Circles of Prometheus

Congratulations to German Gonzalez, a computer engineering student from Columbia, for two wonderful solutions to my own *Circles of Prometheus* puzzle.[23] In the first solution (Fig. 4.28), he used consecutive integers from 1 to 26 at the circle intersections. The eight small circles have the magic sum 62. The two bigger circles have a magic sum of 103.

In the second solution (Fig. 4.29), Gonzalez used consecutive numbers from 1 to 22 and from 54 to 57 to obtain a magic sum of 95 for *all* circles. We do not know if there are other solutions with lesser sums.

Cirri of Euripides

In my *Cirri of Euripides*,[24] the objective is to make the circle and three line sums all the same. So far no one has solved this puzzle. However, Marsha Sisolak, a bilingual kindergarten teacher and mystic from Thousand Oaks, California, found several interesting intermediate solutions using consecutive numbers from 1 to 30.

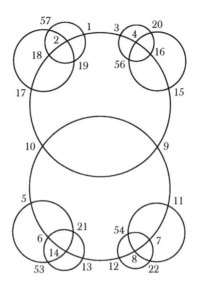

Figure 4.29
Circles of Prometheus.

In Figure 4.30, the straight lines sum to 78. The largest circles sum to 68. The second largest circles sum to 85. The next largest circles sum to 105. The smallest circles sum to 90. Parallel pairs of intersection points produce equal sums on the right and left sides— for example, $18 + 15 = 16 + 17$.

In Figure 4.31, the straight lines sum to 68. The largest circles sum to 95. The second largest circles sum to 109. The smallest circles sum to 82. In this elegant solution, number pairs along the ver-

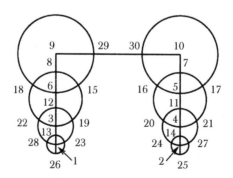

Figure 4.30
Cirri of Euripides.

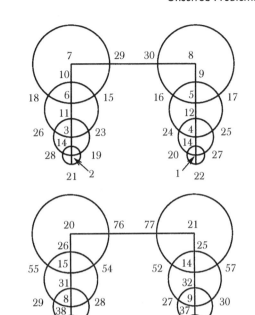

Figure 4.31
Cirri of Euripides.

Figure 4.32
Cirri of Euripides.

tical lines sum to 17, for example, 7 + 10, 6 + 11, 5 + 12, etc. If you sum the vertical pairs of numbers at the circles' intersections, you get a pattern: 16, 14, 16 on the left side and 14, 16, 14 on the right side. The horizontal pairs at the circle intersections sum to the same amount on opposite sides, as in the first solution.

In Figure 4.32, *all* circles sum to 200. The two lines sum to 194. Can you find a smaller magic sum? (Hint: German Gonzalez sent me a solution, using nonconsecutive numbers, with all circle sums of 114 and all three line sums of 194.) Can you find a solution in which the lines have the same sum as the circles?

Unsolved Problems

Various circular patterns are yet to be solved, and I include them here for readers to experiment with (Figs. 4.33–4.38). These are from my Grand Internet Math Challenge,[25] and no one on Earth has been able to find solutions. The goal is to use consecutive numbers at the intersection points to produce magic sums.

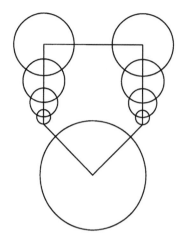

Figure 4.33
Phallic encroachment.

Figure 4.34
Spirogyralia.

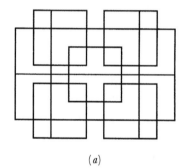

(a)

(b)

Figure 4.35
Lego future.

Figure 4.36
Galadriel rings.

Although no solution has been found for Figure 4.35*a*, Arlin Anderson has discovered the solution in Figure 4.35*b*, which makes use of numbers −17 to 17 (excluding zero). The five small squares, two vertical lines, one center horizontal line, and large rectangle all sum to zero.

Prometheus Unbound

An apt chapter closing is the haunting lines of a beloved poet who used images of circles and spheres in his poems:

> A sphere, which is as many thousand spheres;
> Solid as crystal, yet through all its mass
> Flow, as through empty space, music and light;
> Ten thousand orbs involving and involved.

Purple and azure, white green and golden,
Sphere within sphere; and every space between
Peopled with unimaginable shapes,
Such as ghosts dream dwell in the lampless deep;
Yet each inter-transpicuous; and they whirl
Over each other with a thousand motions,
Upon a thousand sightless axles spinning,
And with the force of self-destroying swiftness,
Intensely, slowly, solemnly, roll on,
Kindling with mingled sounds, and many tones,
Intelligible words and music wild.
With mighty whirl and multitudinous orb
Grinds the bright brook into an azure mist
Of elemental subtlety, like light.

Percy Bysshe Shelley, *Prometheus Unbound*

Figure 4.37
Cleopatra's revenge.

Figure 4.38
Dante's snowflake.

CHAPTER FIVE

Gallery 3: Stars, Hexagons, and Other Beauties

Though the scope of our imagination with all its possibilities be infinite, the results of our construction are definitely determined as soon as we have laid their foundation, and the actual world is simply one realization of the infinite potentialities of being. Its regularities can be unraveled as surely as the harmonic relations of a magic square. A study of magic squares may have no practical application, but an acquaintance with them will certainly prove useful, if it were merely to gain an insight into the fabric of regularities of any kind.

–Paul Carus, in *Magic Squares and Cubes*

God gave us the darkness so we could see the stars.

–Johnny Cash, *"Farmers Almanac"*

This chapter contains an exhibition gallery of my favorite magic stars, hexagons, and other related geometrical constructs.

Simple Magic Stars

A magic star is a variation of a magic square. The numbers are arranged in a star formation such that the sum of the numbers in each of the straight lines formed by the star's points and intersections yields a constant sum.

Figure 5.1 is a magic star containing the numbers 1 through 12, with 7 and 11 omitted. The magic constant for each straight line is 24, which is the smallest possible sum for this range of integers. There is no solution to this particular type of magic star if ten consecutive integers are used.[1]

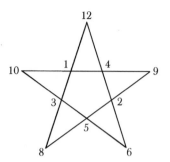

Figure 5.1
Simple magic star.

Figure 5.2 shows magic stars with magic sums of 40, 56, 48, and 32, respectively.[2]

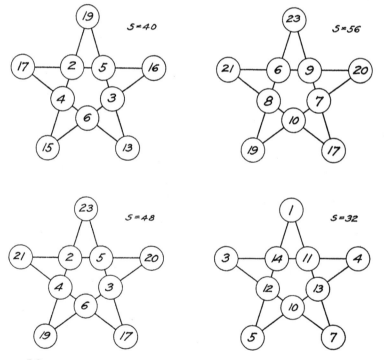

Figure 5.2
Magic stars with magic constants indicated.

Ornate Magic Stars

Figures 5.3 through 5.8 show some magnificent magic stars of sufficient complexity to make the antediluvian gods weep for joy.[3]

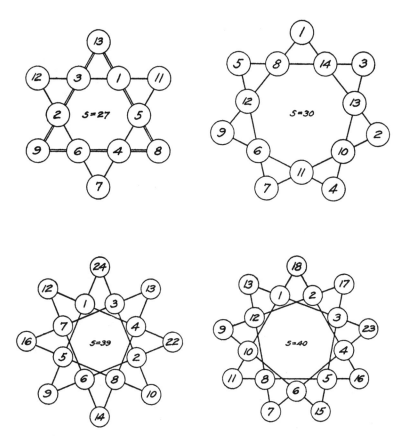

Figure 5.3
Magic stars.

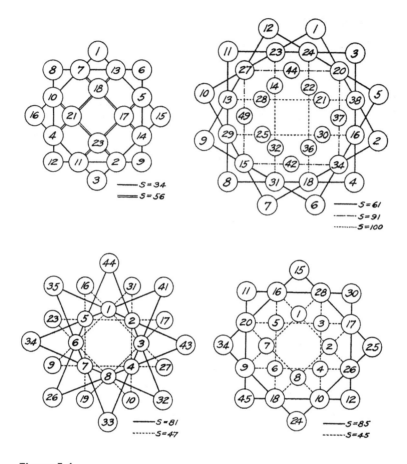

Figure 5.4
Magic stars.

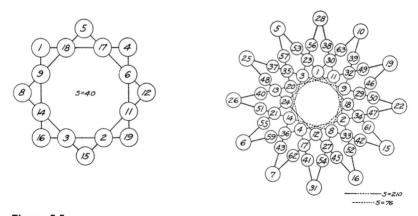

Figure 5.5
Magic stars.

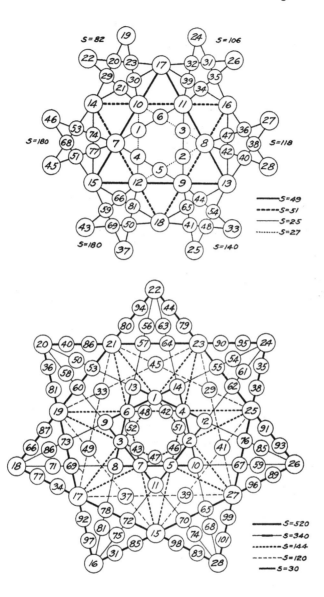

Figure 5.6
Magic stars.

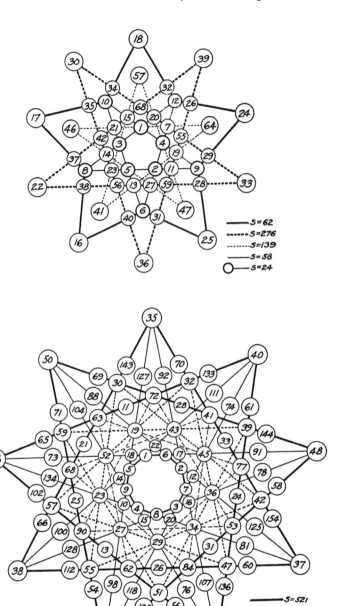

Figure 5.7
Magic stars.

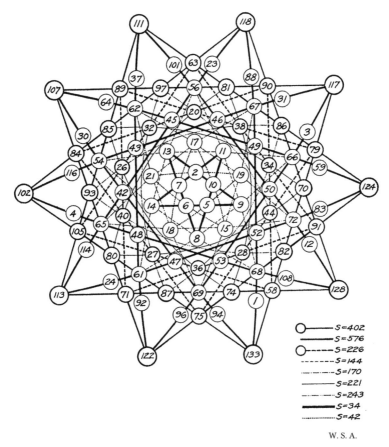

W. S. A.

Figure 5.8
Magic stars.

Magic Hexagonal Tiles

Figure 5.9 is a truly amazing array of numbers.[4] The sum of the integers in any straight line of edge-joined hexagons is 38. Excluding reflections and rotations, this is the only possible magic hexagonal array. There are no solutions for higher orders with more hexagons arranged around the hexagonal cluster in this figure. There is no solution for the next smaller size consisting of seven hexagons. What a lonely pattern this is! (See chapter 2 for more information.)

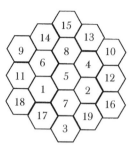

Figure 5.9
Magic hexagonal tiles.

Hexagonal Talisman Tiles

Talisman hexagonal tiles[5] are arrays of numbers from 1 to N^2 in which the difference between any one number and its neighbor is greater than some given constant Φ. In Figure 5.10, $\Phi = 4$.

Can you make talismans for *triangular* arrays of numbers? Can you make an antimagic talisman? What are the minimum and maximum Φ you can create for a hexagonal talisman?

Magic Domino Squares

As explained in chapter 2, domino magic squares contain an arrangement of dominoes so that the fifty-six squares form a 7×7 magic square in which the sum of the pips in each row, column, and diagonal of the square are equal. Figure 5.11 is such a square bordered by a line of seven blank squares. The magic sum is 24.[6]

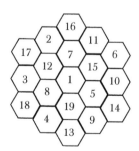

Figure 5.10
Hexagonal talisman tiles.

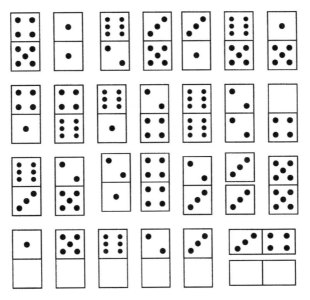

Figure 5.11
Domino magic square.

It is possible to select certain dominoes out of the set and reject the others to make various magic arrangements. Figure 5.12 shows two examples of such magic squares.[7] I've placed a little space between the dominoes to make their arrangements clearer, but you can cram them together to make a perfect overall square.

Magic Interlocked Hexagons

In chapter 2, you saw how hexagons may be packed side-by-side tightly on the surface of a torus and each hexagon divided into six identical triangles (Fig. 2.25). *Magic interlocked hexagons* are figures in which the six triangles around every vertex (an example shown as a circle in Fig. 2.25) contain numbers that sum to 219. The six triangles adjacent to these (forming a starlike shape) also have this sum. Figure 5.13 shows values within the triangles of the interlocked hexagons on a torus.[8]

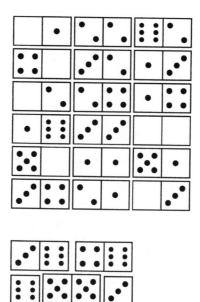

Figure 5.12
Domino magic square.

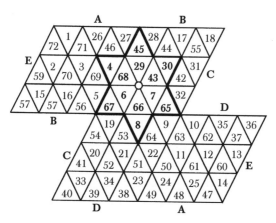

Figure 5.13
Interlocked hexagons on a torus.

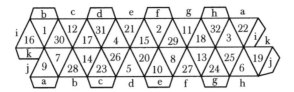

Figure 5.14
A magic rotating ring.

Figure 5.14 shows a special arrangement of the numbers from 1 to 32 on the ring of eight tetrahedra, which is magic in a different sense.[9] (A tetrahedron has four vertices, six edges, and four equivalent equilateral triangular faces.) The four faces of each tetrahedron sum to 66. Corresponding faces, one from each tetrahedron, sum to 132. For example, $9 + 7 + 17 + 31 + 10 + 8 + 18 + 32 = 132$, and so do eight sets of eight faces that wind helically around the ring. For example, $1 + 12 + 31 + 21 + 2 + 11 + 32 + 22 = 132$.

Magic Hexagonal Lattice

Figure 5.15 shows the numbers 1 to 30 arranged so that the corners of each of the nine hexagons sum to 93. There are seven small hexagons and two additional concentric hexagons. These are two of many solutions.[10]

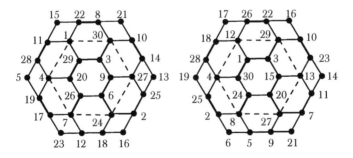

Figure 5.15
Magic hexagons.

More Magic Hexagonal Lattices

Figure 5.16*a* shows a unique solution for a hexagonal lattice arrangement using the integers 1 to 19.[11] The six lines of three numbers, six lines of four numbers, and three lines of five numbers each sum to 38. No other solution for *any* order hexagon is possible. Figure 5.16*b* is formed by picking an arbitrary number, in this case 35, and subtracting each number in Figure 5.16*a* from it. This arrangement has the characteristic that each of the six lines of three integers have the same sum, in this case 67. Each of the six lines of four integers sum to the same value, in this case 102. And each of the three lines of five integers sum the same, in this example, 137.

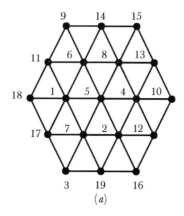

(*a*)

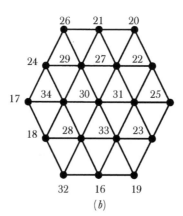

Figure 5.16
Magic hexagons.

(*b*)

Order-8 Magic Square with Embedded Star

9	14	**4**	7	**1**	10	8	15
8	11	5	**10**	16	13	3	2
1	6	**12**	15	**5**	4	14	11
16	**3**	13	2	12	**7**	9	6
16	1	**8**	9	**11**	16	5	2
3	12	13	**6**	8	1	12	13
10	7	**2**	15	**9**	14	7	4
5	14	11	4	6	3	10	15

This order-8 magic square is composed of four order-4 pure magic squares with magic sum 34. The embedded magic star, indicated by bold numbers, is "supermagic" because even the endpoints sum to the constant 34. The diagram at the bottom is just to help you see the embedded star.[12]

Dharmakaya Rings

Figure 5.17 shows the interlocking *Dharmakaya rings* that I challenged colleagues to solve.[13] As part of my "Grand Internet Math Challenge," readers were asked if it were possible to place consecutive numbers from 1 to 15 at the various intersection points and vertices so that the numbers along each circle and line would sum to a constant. Congratulations to Michael Keith (see *Samsara circles*), the first person on Earth to solve the mysteries of my Dharmakaya rings. The magic sum is 40. Additionally, the four points along the triangle edges sum to 40!

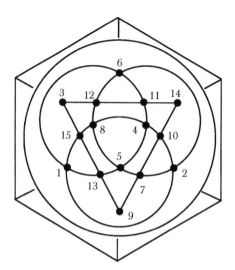

Figure 5.17
Dharmakaya rings.

How many other solutions do the Dharmakaya rings have? Mike suggests that the answer is a walloping 79,824 if we consider all solutions, including those that can be transformed into one another by rotations and reflections. However, it may be more natural to call two numberings the same (isomorphic) if they are equivalent under one of the six-fold symmetries of the figure. In this case, the number of distinct solutions using integers from 1 to 15 is 13,304.

Note that there is a much larger number of possible integer assignments (15! = 1,307,674,368,000) using consecutive integers from 1 to 15, so while 13,304 solutions might seem large, it is a tiny minority of all possible assignments.

Durga Yantra

Yantras are precise geometric forms that have been used for centuries as tools for meditation and self-realization. Each yantra represents a specific quality or aspect of the Divine Creation. In this section, I'd like to apply numbers to the Durga Yantra, which represents "divine radiance unified in a female form."[14] Figure 5.18*a* is a rendering of the Durga Yantra by PennyLea Morris Seferovich.[15]

According to the ancient Hindu scriptures, Durga was the first female divinity in the Universe. This beautiful goddess slew demons and never relied on male protection. I presented the interlocking Durga Yantra on the Internet as a tough challenge for colleagues to solve.[16] Friends were asked if it were possible to place consecutive numbers from 1 to 18 at the various intersection points and vertices so that the numbers along each triangle would sum to a constant. Congratulations to Michael Keith, who finally solved the challenge. The magic sum is 41 (Fig. 5.18*b*).

(*a*)

Figure 5.18
(a) *Durga Yantra. Design by PennyLea Morris Seferovich.* (b) *Magic Durga Yantra (on following page).*

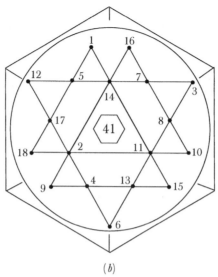

(*b*)

Figure 5.18 (*continued*)

Note 17 provides more information on the goddess Durga for all you budding mystics.

Nirvana Rings

The *Nirvana rings* is another puzzle in my Internet challenge (Fig. 5.19).[18] I asked people to place consecutive numbers from 1 to 13 at the various intersection points and vertices so that the numbers along each triangle and circle would sum to a constant. Congratulations go to Marsha Sisolak for her solution that has a magic sum of 41 with consecutive integers from 3 to 15 for the circles and outer triangle (Fig. 5.19*a*).

If we assign letters to the intersection points, we find that $a + b + c + d + e$ (the sum of a circle) will never equal $a + c + e$ (the sum of a black triangle). Therefore, she did not worry about getting the black triangle vertices to generate the magic sum. Her triangle sums are 32, 34, and 35.

A follow-up challenge was to find a solution for the numbers 1 through 13 such that the *triangle sums* formed consecutive numbers

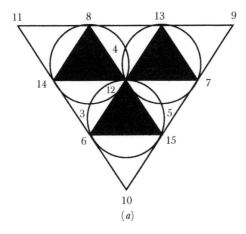

(a)

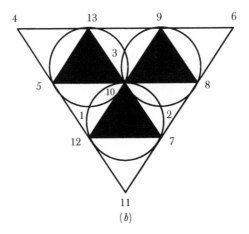

(b)

Figure 5.19
*Nirvana rings. (a) Sisolak solution. (b) Anderson solution. (c)
Miguel solution (on following page).*

while the other sums were magic. Arlin Anderson found the solution in Figure 5.19*b* that produces magic sums of 32 using integers
from 1 to 13. The upper left triangle sum is 28, the upper right triangle sum is 27, and the bottom triangle sum is 29. We believe
there are sixty-five solutions to the Nirvana rings using numbers 1
to 13. Of these solutions, thirty-eight sum to 31 and twenty-seven

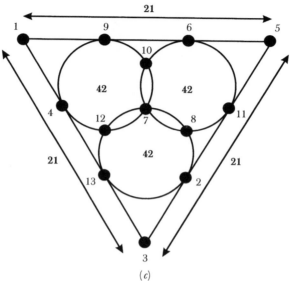

Figure 5.19 (continued)

sum to 32. The sixty-five solutions are increased by a factor of 6 if we include rotations and reflections. There is only one solution with consecutive sums for the black triangles.

In Figure 5.19*c*, Cesar Gomes Miguel, a Brazilian computer scientist, has removed the darkened triangles to produce a figure in which the circles have magic sum S = 42, and the lines have sum $S/2$ = 21, which was Miguel's age at the time of his writing me. He used the consecutive numbers 1 through 13.

I look forward to hearing from readers who have extended this pattern so that it uses more circles and triangles. For example, the triangle may be extended upward so that three more circles are added at the top.

Magic Tetrahedrons

Every magic square of order 4 can be represented as a magic tetrahedron (a solid that has four triangular faces).[19] The tetrahedron is

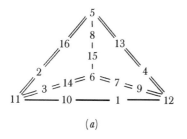 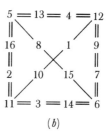

(a)   (b)

Figure 5.20
Magic tetrahedron (a) and corresponding square (b).

magic because the numbers along its edges have the same sum, for example, $11 + 2 + 16 + 5 = 5 + 13 + 4 + 12$. The double lines in the magic square (Fig. 5.20b) correspond to the double lines in a magic tetrahedron (Fig. 5.20a). The diagonal single lines in Figure 5.20b correspond to the other two edges of the tetrahedron in Figure 5.20a.

Window Ornaments

Boris Kordemsky, a Russian high school mathematics teacher born in 1907, calls magic geometries comprised of star and circles "window ornaments."[20] The problem is as follows. A store owner selling semiprecious stones uses a five-pointed star made of circular spots held together by wire. The fifteen spots hold one through fifteen stones (each number used once). Each of the five large circles holds forty stones, and there is a total of forty stones at the five ends of the star. Figure 5.21 shows one solution. Are there others?

Planetariums

Figure 5.22 shows a small "planetarium" on the left in which there are four planets in each orbit and four along each radius. Weights of the planets in the small planetarium are expressed by integers from 1 through 16. In the large planetarium, on the right, the weights go from 1 through 25. One fascinating puzzle, probably

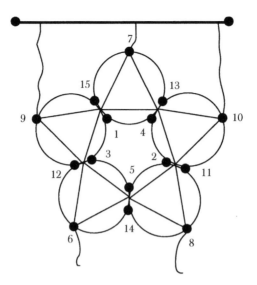

Figure 5.21
Window ornament.

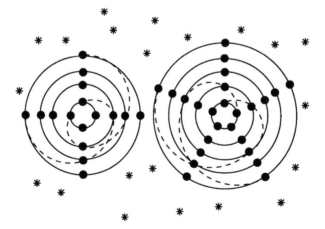

Figure 5.22
Planetariums.

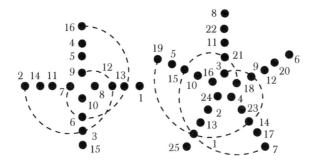

Figure 5.23
Planetariums.

developed in the 1950s,[21] is as follows. Your task is to arrange the weights in Figure 5.22 so that each planetary system is in equilibrium, with sums of 34 for the small planetarium and 65 for the large. The following interesting criteria must be met:

1. The weights along each radius must add up to the magic sum.
2. The weights around each orbit must add up to the magic sum.
3. The weights spiraling in from the outside orbit to the inside orbit in both directions (see dashed lines in Fig. 5.22) must add up to the magic sum.
4. In the small planetarium only, the four weights in each adjacent pair of orbits of each adjacent pair of arms must sum to the magic constant.

Figure 5.23 shows one solution for each planetarium, but the spiral sums are not correct. Are there solutions with the correct spiral sums? I would be interested in seeing new planetariums created by readers.

Triangle Overdrive

The rectangular ornament in Figure 5.24 consists of sixteen small triangles that contain the integers 1 through 16.[22] If you look closely, you'll see there are six larger, overlapping right triangles formed from these small triangles. Figure 5.24 shows one way that you can

Figure 5.24
Triangle overdrive.

make the sums of integers in each of these large triangles 34. Are there other solutions? I would be interested in hearing from readers who have created larger ornaments with similar properties.

The Trapezoid Catastrophe

You are carrying an expensive piece of colored, triangular glass that accidentally slips from your fingers and develops hairline cracks, which themselves are triangular. Figure 5.25 shows your cracked large triangle with three overlapping triangles, each containing four cells, and three trapezoids, each with five cells.[23] The cells contain numbers 1 through 9 so that each triangle adds up to 17 and each trapezoid to 28. It is possible to find six arrangements of this figure with triangle sums of 20 and trapezoid sums of 25, and two arrangements with triangle sums of 23 and trapezoids of 22 (Fig. 5.26). Are there other solutions? I would be interested in hearing from readers who generate similar figures containing more triangles.

Lotus Flower from Ganymede

I posed the beautiful and enigmatic Lotus Flower from Ganymede to top mathematicians and scientists in 1999.[24] Their goal was to fill the flower in Figure 5.27 with numbers to produce

Figure 5.25
The trapezoid catastrophe.

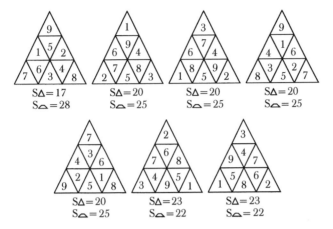

Figure 5.26
The trapezoid catastrophe.

Figure 5.27
Ganymede flower.

magic spirals. Notice that each petal has two spirals that intersect it. (For example, the petal with 90 near the flower's bottom center sits on two spirals, one composed of the numbers 121, 104, 90, 71,

57, 40, 26, and 7 and one composed of the numbers 119, 105, 90, 72, 59, 37, 22, and 12. The challenge was to place a different integer in each petal so that each spiral in the flower has the same sum.

Joseph DeVincentis, a chemical engineer and Zen master from Winthrop, Massachusetts, was able to arrive at a solution, filling the flower with numbers 1 to 128 so that each spiral sums to 516. The following table represents the uncurled flower. Each row in turn represents a single concentric layer of petals; each diagonal represents one of the spirals. The first numbers in each row represent one of the spirals (1, 32, 46, 51, 65, 96, 110, and 115). To form the spirals that go in the opposite direction, you can examine a column (for example, 8, 26, 41, 55, 68, 94, 109, and 115). Figure 5.27 should make this more obvious.

1	2	3	4	5	6	7	8	9	10	11	12	13	14	15	16							
	32	31	30	29	28	27	26	25	24	23	22	21	20	19	18	17						
		46	45	44	43	42	41	40	39	38	37	36	35	34	33	48	47					
			51	52	53	54	55	56	57	58	59	60	61	62	63	64	49	50				
				65	66	67	68	69	70	71	72	73	74	75	76	77	78	79	80			
					96	95	94	93	92	91	90	89	88	87	86	85	84	83	82	81		
						110	109	108	107	106	105	104	103	102	101	100	99	98	97	112	111	
							115	116	117	118	119	120	121	122	123	124	125	126	127	128	113	114

Uncurled flower

To solve this flower, Joseph used brute-force thinking and testing, assisted occasionally by a computer. We do not know how many solutions the Ganymede flower has, and we are currently looking for additional magical properties. As one colleague told me, "I could spend a lifetime studying this flower and never fully glean all its properties and wonders."

Here is what we do know. Using Joseph's solution, we believe it is possible to create fifteen other solutions by rotating the inner four concentric layers with respect to the outer four. We can create another sixteen solutions using the direction of four layers (e.g., start the fifth layer running in the opposite direction around the flower; the last three will likewise reverse). Also, each of those solutions is only one of a separate family of 8! solutions that differ in the order that the multiples of 16 were added to the different layers (e.g., add 16 to the first layer and leave the second layer unmodified, instead of adding the 16 to the second layer). Finally, we believe that there are two more degrees of symmetry that can be expressed by negating (subtracting from 17) the numbers on either the inner four layers or the outer ones. We believe swapping one of the sets of four layers (swapping the first and fourth layers, and the second and third) achieves the same effect. There may be a total of $16 \times 2 \times 40{,}320 \times 4 = 5{,}160{,}960$ solutions based on this one pattern, not counting reflections or rotations of the whole figure.

Magic Spider

Using the magic spider configuration in Figure 5.28, your goal is to label the nodes (dots) with integers so that a magic constant is achieved for the head (a, b, d, c) and five closed figures that make up the body. Different integers should be used at each node. If possible, the integers should be consecutive. The problem is made interesting by the various interweavings of body layers.

One solution has a magic constant of 80 and can be created by assigning $a = 10$, $b = 7$, $c = 3$, $d = 60$, $\beta = 18$, $\chi = 20$, $z = 5$, $w = 5$, $n = 6$, and all the rest of the nodes equal to 1. This solution is not very elegant due to the repetition of 1s. Can you arrive at something more worthy of a powerful arachnid god?

A Pandiagonal Torus

Figure 5.29, by magic square guru Harvey Heinz, is a torus projected onto a plane, and it may be used as an order-5 pandiagonal

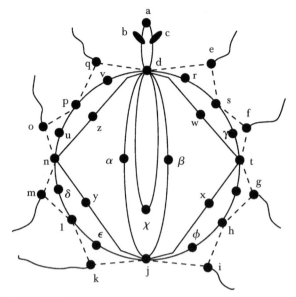

Figure 5.28
Magic spider.

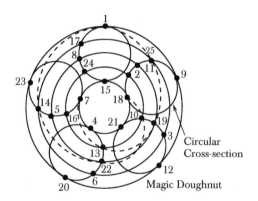

Figure 5.29
Pandiagonal torus.

magic square generator.[25] For example, consider the following two magic squares labeled *A* and *B*. Next, examine the torus. Start at number 1 and follow the big circles clockwise to generate the rows of the *A* magic square. Start at number 2 and follow the big circles

1	9	12	20	23
17	25	3	6	14
8	11	19	22	5
24	2	10	13	16
15	18	21	4	7

A

2	11	25	9	18
10	19	3	12	21
13	22	6	20	4
16	5	14	23	7
24	8	17	1	15

B

to generate the columns of the *B* magic square. The big circles, little circles, or spirals in either direction may be used to represent rows, columns, or diagonals. For example, there are five concentric rings (e.g., 1, 9, 12, 20, 23, and back to 1), five cross-sectional rings (e.g., 1, 15, 24, 8, 17, and back to 1), five clockwise spirals (e.g., 1, 25, 19, 13, 7, and back to 1), and five counterclockwise spirals (e.g., 1, 14, 22, 10, 18, and back to 1). Any one set of these closed paths may be used to generate the rows or the columns of an order-5 pandiagonal magic square. For diagrammatic clarity, only two of the spiral paths are shown in dashed lines.

Fractals Reflected in Water

Figure 5.30 uses consecutive numbers from 1 to 17.[26] The fractal triangles above the water line and below the water line each have 85 for a magic sum. (The 17 is shared by the above-water and below-water set of triangles.) The sums of the *squares* of the numbers in the above-water and below-water triangles are both 1037. And if that isn't sufficiently miraculous for you—the sums of the *cubes* of the numbers in the above-water and below-water triangles are both 14,161. In principle, you could keep adding smaller triangles to the figure, but I do not know if magic patterns can be formed for deeper levels of nestings. One reason this figure may be considered fractal is that it can exhibit similar nested triangular structures for a range of size scales.

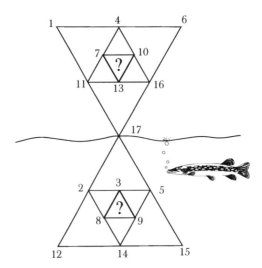

Figure 5.30
Fractals reflected in water.

Prime Star

Figure 5.31 is a magic pentagram containing a set of consecutive prime numbers that have the lowest possible values.[27] The lowest prime number in the star, 13,907, is the 1644th prime number. There are twelve basic solutions to the order-5 consecutive primes magic star.

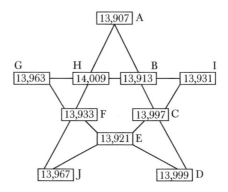

Figure 5.31
Consecutive primes magic star.

#	A	B	C	D	E	F	G	H	I	J	Sum
1	13907	13913	13997	13999	13921	13933	13963	14009	13931	13967	55816
2	13907	13913	13999	13997	13921	13967	13931	14009	13963	13933	55816
3	13907	13933	13967	14009	13931	13913	13963	13999	13921	13997	55816
4	13907	13933	14009	13967	13931	13997	13921	13999	13963	13913	55816
5	13907	13967	13933	14009	13963	13913	13931	13997	13921	13999	55816
6	13907	13967	14009	13933	13963	13999	13921	13997	13931	13913	55816
7	13913	13907	13997	13999	13921	13963	13933	14009	13967	13931	55816
8	13913	13907	13999	13997	13921	13931	13967	14009	13933	13963	55816
9	13913	13931	13963	14009	13933	13907	13967	13997	13921	13999	55816
10	13913	13963	13931	14009	13967	13907	13933	13999	13921	13997	55816
11	13931	13921	13967	13997	13907	13913	13999	13963	13933	14009	55816
12	13963	13921	13933	13999	13907	13913	13997	13931	13967	14009	55816

Unsolved Problems

Various more complex patterns are yet to be solved, and I include them here for readers to experiment with (Figs. 5.32–5.36). These are from my Grand Internet Math Challenge,[28] and no one on Earth has been able to find solutions. The goal is to use consecutive numbers at the intersection points to produce magic sums. I look forward to hearing from readers who find partial solutions to these problems or prove them impossible to solve.

Figure 5.32 shows a hypercheckerboard. Here are the rules you need to solve it. First, look at the small squares. Your mission is to place an integer in each of these squares. The sums of the integers forming a diagonal must always be the same. If this figure appears too complex, feel free to truncate the figure as you wish in order to remain sane.

Figure 5.32
Hypercheckerboard.

There is another set of rules that may be applied to the same figure. You see two kinds of cells. Cells *A* are the small squares that we just mentioned. Cells *B* are larger polygons with a staircase edge. Place an integer in each of the Cells *A* so that the sum of these integers adds up to the number in each Cell *B*. Be creative. If you don't like these rules, feel free to alter them to create a magic figure.

Figure 5.33
Islamic magic geometry.

In Figure 5.33, each central star must contain an integer that is equal to the sum of the six numbers in each surrounding hexagon. Truncate the figure as you wish in order to remain sane.

In Figures 5.34 through 5.36 from the Dover Pictorial Archive, your goal is to place consecutive integers into the various cells to

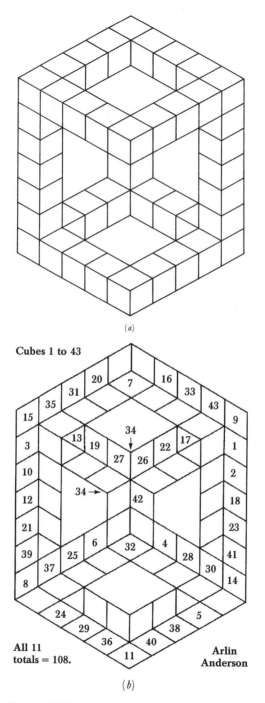

(a)

(b)

Figure 5.34
Hyperdimensional twisted frame. (a) Naked frame.
(b) Magic cubic treatment with sum 108.

create magic sums in as many directions as possible. Interestingly, Arlin Anderson was able to treat Figure 5.34 as a set of cubes and create a magic labeling for the cubes (not the faces of the cubes). Arlin used consecutive numbers starting at 1. The cubes sum to 108 in all directions. For example, the top left set of cubes corresponds to $15 + 35 + 31 + 20 + 7 = 108$. Even the small crosslink at center produces a magic sum $(32 + 42 + 34)$!

Figure 5.35
Unraveled.

Figure 5.37 shows Arlin Anderson's solution to the cryonic cube. The goal was to create a numbering of the visible faces such that the rows and columns would produce a magic sum. Arlin was able to use consecutive numbers starting from 1 to create such a numbering. Here are a few example sums: $1 + 17 + 24 + 18 + 22 + 2 = 84$, $24 + 5 + 25 + 8 + 10 + 12 = 84$, and $2 + 13 + 27 + 12 + 7 + 23 = 84$. An unsolved challenge would be to incorporate more of the diagonals in this figure in the magic sum, or to extend this figure by adding a $4 + 4$ array of cubes to the base.

Figure 5.36
Bicubic mystery.

Figure 5.37
Cryonic cube.

The Nagle Yantra (Fig. 5.38) designed by Nancy Nagle[29] was solved in 1999 by Arlin Anderson, who was able to place consecutive numbers from 1 through 30 at the triangles' intersections and vertices. In Figure 5.38, the magic sum is 85. For example, $1 + 2 + 24 + 28 + 30 = 85$. We believe it is possible to create many other labelings that result in a magic sum. No one has been able to create a nontrivial multiplication magic figure so that products along each line segment are equal. This is an unsolved challenge.

The Great Fubine

In January 2000, I came across the energetic octogenarian James Nesi from New York City, who told me about a wonderful

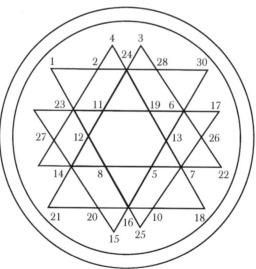

Figure 5.38
Nagle Yantra. (Top illustration by Nancy Nagle.)

collection of magic figures designed by the late, great Fubine. Fubine, whose real name was Cipriano Ferraris, died at the age of 72 on October 30, 1958. Most of his fantastic puzzles were created in the 1930s and decorated the walls of the National Puzzler's League in New York City. Fubine's designs ranged from simple squares through a wide variety of linear geometric shapes and three-dimensional figures. Rows, columns, spokes, and diameters consisted of lines of numbers, no single one of which was repeated and whose totals were always the same. One of his great works was on display at the General Motors exhibit at the 1939 World's Fair in New York.

James Nesi knew Fubine well. Nesi writes,

> Fubine's admirers had little to do but look at his creations with astonishment. Praise for Fubine's accomplishments was not unanimous, and I recall niggardly compliments he received from some of the mathematicians who did not consider his work particularly remarkable. I do not recall, however, that anyone ever tried to match his accomplishments, but try or not, no one has ever succeeded. Moreover, few on either side ever realized that the designs themselves, both in conception and realizations, were truly works of art regardless of the mathematics involved.[30]

Fubine had many ups and downs in his life. In 1929, he lost all his money in the great stock market crash. He found himself in near suicidal state and distracted himself by creating ever-larger magic squares. As time passed, his thoughts of suicide were supplanted with an obsession for magic configurations. "Magic squares saved my life," Fubine once confessed.

Today most of Fubine's works are lost, but James Nesi sent me some of the remaining designs to save for posterity in this book. Figure 5.39 is a portion of an elaborate work dedicating New York's Leonardo da Vinci Art School on the quincentennial of the Italian master's birth. Crammed with numbers are two stars and a zigzagging border with dozens of lines totaling either 1452 or 1952. Surrounding a sketch of da Vinci is the artist's famous quote:

Figure 5.39
A portion of Fubine's da Vinci configurations.

"Beauty perishes in life, not in art." Figure 5.40 is a magnification of a portion of Figure 5.39, which itself is only a portion of the larger work created in 1935.

Figure 5.41 is another Fubine creation, a portion of which is magnified in Figure 5.42. The square is dedicated to Benjamin Franklin and was created in 1936. Although very difficult to see at this magnification and location in the square, the numbers in the two borders are consecutive from 139 to 546 and from 589 to 756. In the inner border there are eighty-four circles containing four numbers each and thousands of geometrical combinations that total 1706. In the outside border there are eighty-four 2×2 and millions of other geometrical combinations totaling 1790. The center 25×25 square is an arrangement that Benjamin Franklin

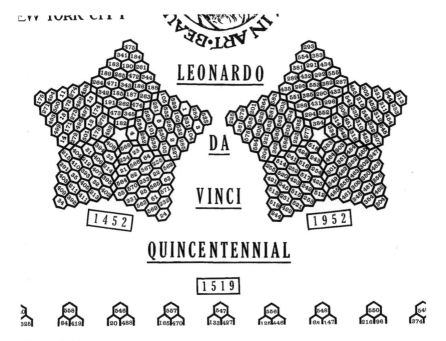

Figure 5.40
A magnification of part of Figure 5.39.

would have enjoyed because it produces two 15×15 magic squares, fifty 5×5 magic squares, and one hundred 3×3 magic squares.

Figure 5.43 is a fragment of another Fubine construction, too complex to describe succinctly, and dedicated to Alexei Stakhanov, a famous Soviet coal miner who developed quick assembly line techniques to expedite Soviet war efforts. The magical configuration was created in 1936, and Figure 5.44 shows a magnification of part of the original configuration. Fubine described this by saying,

> The numbers represent days. The design produces a series of totals of three numbers such that when added together and the grand total divided by a certain number of years, the answer will always be 365 days and 6 hours. For example, it produces 726 totals of 2190, 726 of 2195, 729 of 2192, and 729 totals of 2191 days. By adding these four different totals

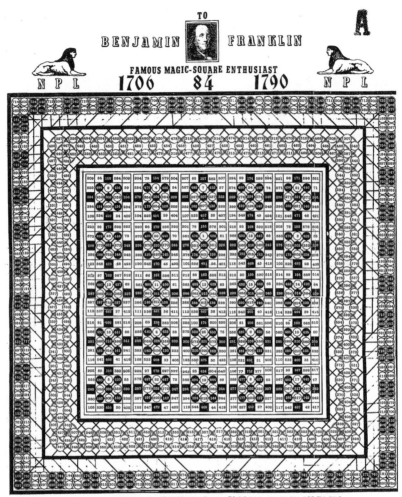

Figure 5.41
A portion of Fubine's Franklin configurations.

together we get 8766 days which, when divided by 24 hours, equal 365 days and 6 hours.

Readers may write to me for further information on this "numerical constellation."

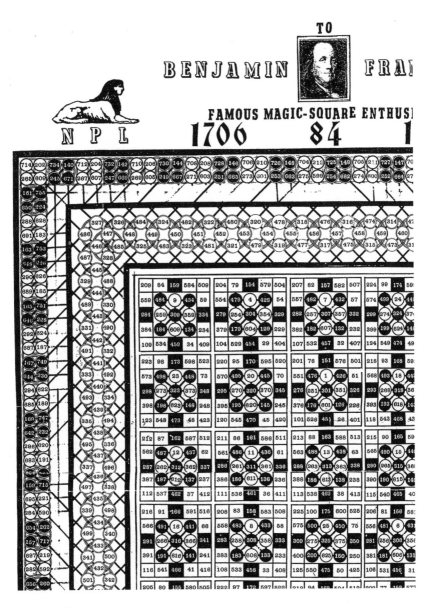

Figure 5.42
A magnification of part of Figure 5.41.

Figure 5.45 is a Fubine square from 1940. It uses the consecutive numbers 1 to 256. As a whole, it is a perfect magic square, totaling 2056 in all directions. Here are some other amazing properties:

1. If you subtract the total of the numbers in circles from the total of the numbers in squares, and divide by the total of the numbers in hexagons, you get 1 in all directions.
2. If you subtract the total of the numbers in the circles and hexagons from the total of the numbers in the squares, the sum will be 0 in all directions.
3. In the overall square are nine 8×8 perfect magic squares, each of which produces the same features just described relating to hexagon, circle, and square sums.

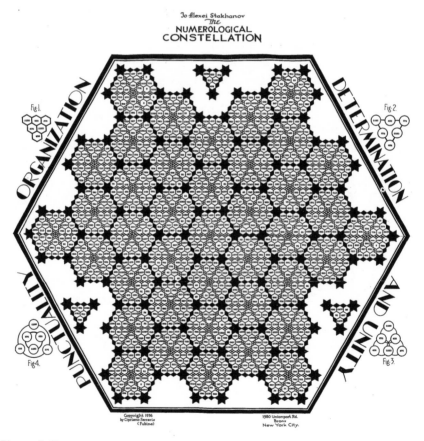

Figure 5.43
A portion of Fubine's "Numerical Constellation."

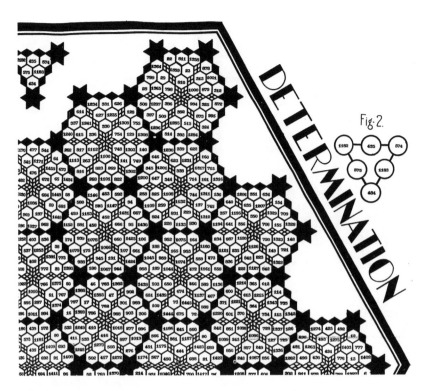

Figure 5.44
A magnification of part of Figure 5.43.

4. There are also sixteen 4×4 and four 12×12 magic squares, each of which has the same features just described relating to hexagon, circle, and square sums.

5. As a whole, by adding the numbers in the little squares of each column, the total will be 1028 in all directions. The same result is also obtained by adding the numbers in the circles and hexagons.

6. The nine 8×8 squares also have the same features just described, namely, the total will be 514 in all directions, and the same result is also obtained by adding the numbers in circles and hexagons.

It is possible that Fubine's entire collection of magic configurations exists somewhere, perhaps cluttering an ancient attic. James

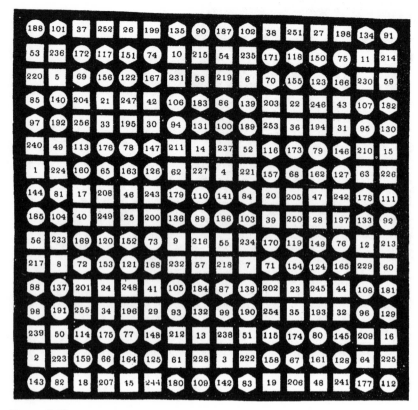

Figure 5.45
Fubine's special magic square.

Nesi has tried for a long time to contact possible relatives, but so far he has been unsuccessful.

Some Final Thoughts

There is no science that teaches the harmonies of nature more clearly than mathematics, and the magic squares are like a mirror which reflects the symmetry of the divine norm immanent in all things, in the immeasurable immensity of the cosmos and in the construction of the atom not less than in the mysterious depths of the human mind.

–Paul Carus, in W. S. Andrews's *Magic Squares and Cubes*

Picture in your mind an infinite floor tiled by adjacent marble slabs, one foot on a side. On each tile is a number, and each 12×12 tile set forms a pandiagonal square. As you learned in this book, each pandiagonal square forms a continuous tiling of overlapping magic squares. With each step you take along the floor, you feel as if you are in an ever-changing magic square sparkling in and out of existence as you travel. No matter in which direction you travel, you feel whole, complete, surrounded by patterns stretching for as far as your eye can see. It's an awesome vision, isn't it?

We are truly lucky we can contemplate magic squares. I know that this statement seems a bit strange, but consider the following. We can hardly imagine a mountain gorilla understanding the significance of magic squares, yet the gorilla's genetic makeup differs from ours by only a few percentage points. In this book's Introduction, we discussed the simple Lo-Shu magic square. Can you imagine trying to teach the significance of this to an orangutan?

Yet, the tiny genetic differences between us and apes produce differences in our brains. If we could somehow further alter our brains, we could contemplate a variety of magic squares and other geometries to which we are now totally closed. What magical

4	9	2
3	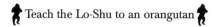	7
8	1	6

Teach the Lo-Shu to an orangutan

squares, tesseracts, and stars are lurking out there that we can never understand? What new aspects of reality could we absorb with extra cerebrum tissue? What exotic magic squares could swim within the additional folds? Philosophers have suggested that the human mind is unable to find answers to some of the most important questions, but these same philosophers rarely thought that our inability was due to an organic deficiency shielding our psyches from higher knowledge.

Certain male wasps, bees, and flies are so in tune with a particular flower's geometry that they cannot help but mate with it. Entomologists even have a name for this practice: *pseudocopulation.* If the Yucca moth, with only a few ganglia for its brain, can recognize the geometry of the yucca flower from birth, how much of our ability to contemplate magic cubes and tesseracts is hardwired into our convolutions of cortex? Our mathematical capacity is a function of our brain. There is an organic limit to our mathematical depth.

It is notable that magic squares and their kin are constructed using integers, and there are still many as yet unsolved problems relating to magic figures. The brilliant mathematician Paul Erdös could pose problems using integers that were often simple to state but notoriously difficult to solve. Erdös believed that if one can state a problem in mathematics that is unsolved and over one hundred years old, it is a problem dealing with integers. Gazing at all the integer arrays in this book reminds me that there is a harmony in the universe that can be expressed by whole numbers. Numerical patterns describe the arrangement of florets in a daisy, the reproduction of rabbits, the orbit of the planets, the harmonies of music, and the relationships between elements in the periodic table. Several times in this book I've alluded to potential practical uses for magic squares, although this avenue is ripe for future exploration. Throughout history, experiments originating in the play of the mind have found striking and unexpected practical applications. I predict that additional applications will be found for magic squares in the twenty-first century.

Ancient people, such as the Greeks and Chinese, had a deep fascination with numbers. Could it be that in difficult times numbers were the only constant thing in an ever-shifting world? To the Pythagoreans, an ancient Greek sect, numbers were tangible, immutable, comfortable, eternal–more reliable then friends, less threatening than Zeus. Maybe the appeal of magic squares is their simple, constant rules that nevertheless produce awesome patterns. Today some Moslems, like the Druze in the Middle East, use simple mathematics as a prelude to contemplation of God in order to wean their minds away from Earthly things to a more abstract mode of thought.

Many researchers through history have noted the prevalence of integer patterns in geometry and mysticism. Perhaps there is something special about integers in the fabric of the universe. Leopold Kronecker (1823–1891), a German algebraist and number theorist, once said, "The primary source of all mathematics are the integers." Since the time of Pythagoras, the role of integer ratios in musical scales has been widely appreciated. In the eighteenth century, French chemist Antoine Lavoisier discovered that chemical compounds are composed of fixed proportions of elements corresponding to the ratio of small integers. This was very strong evidence for the existence of atoms. In 1925, certain integer relations between the wavelengths of spectral lines emitted by excited atoms gave early clues to the structure of atoms. The near-integer ratios of atomic weights was evidence that the atomic nucleus is made up of an integer number of similar nucleons (protons and neutrons). The deviations from integer ratios led to the discovery of elemental isotopes (variants with nearly identical chemical behavior but with different radioactive properties). Small divergences in the atomic weight of pure isotopes from exact integers confirmed Einstein's famous equation $E = mc^2$ and also the possibility of atomic bombs. Integers are everywhere in atomic physics. Integer relations are fundamental strands in the mathematical weave–or as German mathematician Carl Friedrich Gauss said, "Mathematics is the queen of sciences–and number theory is the queen of mathematics."

I side with Martin Gardner, who I interpret as saying that nature is almost always describable by simple formulas not because we have invented mathematics to do so but because of some hidden mathematical aspect of nature itself. For example, Martin Gardner writes in his 1985 work *Order and Surprise*:[1]

> If the cosmos were suddenly frozen, and all movement ceased, a survey of its structure would not reveal a random distribution of parts. Simple geometrical patterns, for example, would be found in profusion–from the spirals of galaxies to the hexagonal shapes of snow crystals. Set the clockwork going, and its parts move rhythmically to laws that often can be expressed by equations of surprising simplicity. And there is no logical or a priori reason why these things should be so.

I like to imagine walking through some futuristic museum of math in which magic squares in all dimensions are exhibited, somehow laid bare before my feeble human eyes. The museum contains every magic geometry. To my left is the Hall of Magic Tesseracts, to my right the Hall of Hyperspheres. In the distance are structures so alien that they defy description. I imagine the tesseracts chirping, the hyperspheres humming–a whole community of patterns existing in a crystal continuum. Even if our minds have difficulty understanding all the structures, we would realize that wonders such as these are just within our grasp, separated from us by the filmiest of veils. Yet, despite our limitations, we need not despair, because computers will be our prosthetics and crutches allowing us to explore a seemingly impenetrable new universe of structures. Our heirs, whatever or whomever they may be, will explore magic squares to degrees we cannot currently fathom. They will discover that magic squares are a symphony of patterns played in many keys. There are infinite harmonies to explore.

Zen Buddhism encourages meditation whereby one purges oneself of emotion and achieves a stillness at one's core. An ardent magic-square explorer is a "mathematical samurai," someone who trains to be neither fearful of defeat nor hopeful of victory and thus enters combat with a neutral attentiveness, indifferent to–but pre-pared for–the difficult demands of each instant. I suppose that there might be a small minority of mathematical scholars who occasionally look down upon the recreational mathematicians who dabble with magic squares.[2] However, many of the mathematical samurai respond by teaching themselves mathematical arts and can sometimes surpass the seasoned mathematicians. The mathematical exhibitions in this book–the tesseracts, circles, and stars–are perfect examples of the mathematical samurai's accomplishments and devotion to the Zenlike quest. Of course the magic square is not the pinnacle of mathematical thought–it's just one of many jewels within the treasure chest of mathematics. Many different kinds of structures and patterns, from all branches of mathematics, can energize the sprit and permit years of contemplation.

I have found that while studying and doing research on magic squares, sitting in my home office or in a library or while gazing at the computer screen, I often get a flicker of happiness, or dare I say "transcendence" or "wonder," that seems to bring the magic square to life, and then I'm busy transcribing the squares for this book, back in a more academic mode of thought, but, for a few seconds, I felt touched and mystically elevated. It's hard to predict or plan for these moments or to determine which magic square will do the trick. My Jewish friends tell me that this is like a Kabbalistic flash that Jews get when studying the torah, or Kabbala, or reciting sacred words and debating their meaning.

To borrow a concept from Robert Pirisig, author of *Zen and the Art of Motorcycle Maintenance,* the Buddha, the godhead, resides quite as comfortably in the patterns of a magic square as he does within temple walls or in the petals of a flower; to think otherwise is to demean the Buddha–which is to demean oneself. The path to new magic structures in the twenty-first century is long. The exhibition in this book is incomplete–just a tiny zoo. It's as if I've taken a spoon of sand from the infinite landscape of awareness around us and called that spoon of sand the world.

Notes

1. I found this quote on my desk. It was written in a nearly illegible scrawl. I suspect that the cited individual may be a pseudonym for whoever left me the mysterious quote.

2. Ibid.

Preface

1. Quote from a speech by Benoit Mandelbrot, *Conference on a New Space for Culture and Society* (*New Ideas in Science and Art*), Prague Castle, Czech Republic, November 19–23, 1996 (for more information, see http://pconf.terminal.cz/participants/mandelbrot.html).

2. Other mathematicians disagree with my philosophy and believe that mathematics is a marvelous invention of the human mind. One colleague uses poetry as an analogy. He writes to me

> Did Shakespeare "discover" his sonnets? Surely all finite sequences of English words "exist," and Shakespeare simply chose a few that he liked. I think most people would find the argument incorrect and hold that Shakespeare *created* his sonnets. In the same way, mathematicians create their concepts, theorems, and proofs. Just as not all sequences of words are sonnets, not all grammatical sentences are theorems. But theorems are human creations no less than sonnets.

3. Roger Penrose, *The Emperor's New Mind: Concerning Computers, Minds, and the Laws of Physics* (New York: Penguin, 1991).

4. Ibid.

5. A colleague suggested that the magic squares in this book don't even begin to compare in mathematical depth and philosophical wonder with other mathematical concepts such as Cantor's theorem on the uncountability of the real numbers. I tackle this sort of subject in my book *Keys to Infinity*.

6. One reviewer for this book suggested that the reference to Timothy Leary is very much dated and that most young readers will not relate to it. For those of you not familiar with Timothy Leary, see www.leary.com. In the 1960s, Leary experimented with psychedelics such as LSD. Another colleague suggested perhaps a more poetic metaphor: Arithmetic satori is like buying an old chest at a thrift store, only to find an original oil painting by Jan Vermeer beneath the false bottom.

7. Benjamin Franklin, "Letters and Papers on Philosophical Subjects by Benjamin Franklin, L.L.D., F.R.S," printed in London (1769).

8. Ibid.

9. W. S. Andrews, *Magic Squares and Cubes* (1917; reprint, New York: Dover, 1960).

Acknowledgments

1. Bhairava Yamala, as quoted by Titriana Leucci in "The Megalithic Anthropomorphic Stone-Slabs from Tamil Nadu: A Comparative Analysis with the Prehistoric Mediterranean Mother Goddess Images," a paper presented by archaeologist Leucci at the *Eighth International Tamil Conference* held in January 1995 at Tangevore, Tamil Nadu, India.

Introduction

1. Joseph S. Madachy, *Madachy's Mathematical Recreations* (New York: Dover, 1979), 87.

2. Dame Kathleen Ollerenshaw and Sir Hermann Bondi, "Magic Squares of Fourth Order," *Philosophical Transactions of the Royal Society, London, Series A*. 306:443–532 (1982).

3. Martin Gardner, *Time Travel and Other Mathematical Bewilderments* (New York: Freeman, 1987), 216.

4. Ibid.

5. Joseph S. Madachy, *Madachy's Mathematical Recreations*, 86.

6. John Lee Fults, *Magic Squares* (La Salle, Ill.: Open Court, 1974), 4.

7. Keith Ellis, *Number Power* (New York: St. Martin's Press, 1978).

8. Annemarie Schimmel, *The Mystery of Numbers* (New York: Oxford University Press, 1993).

9. Ibid.

10. Keith Ellis, *Number Power*, 205.

11. Clifford Pickover, *The Loom of God* (New York: Plenum, 1997).

12. Henry Dudeney, *Amusements in Mathematics* (1917; reprint, New York: Dover, 1970), 119–126.

13. Ibid.

14. W. S. Andrews, *Magic Squares and Cubes*, 19.

15. Ibid., 27.

16. *Hermetica* are works of revelation dealing with occult, theological, and philosophical subjects attributed to the Egyptian god Thoth (Greek Hermes Trismegistos, i.e., "Hermes the Thrice-Greatest"), who was said to be the inventor of writing and all the arts that depend on writing. During Hellenistic times, there was an increasing distrust of traditional Greek rationalism and destruction of the line between science and religion. Hermes-Thoth was one of several gods to whom humans turned for wisdom.

17. Frances Yates, *Giordano Bruno and the Hermetic Tradition* (Chicago: University of Chicago Press, 1964).

Chapter 1. Magic Construction

1. Marvin Kaye, *The Handbook of Mental Magic* (New York: Stein and Day, 1975), 239.

2. See, for example, John Lee Fults, *Magic Squares*, and W. S. Andrews, *Magic Squares and Cubes*.

3. Jan Gullberg, *Mathematics: From the Birth of Numbers* (New York: W. W. Norton, 1997), 212.

4. John Lee Fults, *Magic Squares*, 18.

5. Ibid.

6. W. S. Andrews, *Magic Squares and Cubes*, 5.

7. Eric Weisstein, *CRC Concise Encyclopedia of Mathematics* (New York: CRC Press, 1998), 1128.

8. W. S. Andrews, *Magic Squares and Cubes*, 244–245.

9. John Hendricks, "The Diagonal Rule for Magic Cubes of Odd Order," *Journal of Recreational Mathematics* 20(4):192–195 (1988).

10. W. W. Ball and H. S. M. Coxeter, *Mathematical Recreations and Essays* (New York: Dover, 1987), 206–210.

11. John Hendricks, personal communication, and "Hruska's Conjecture" (pamphlet), Victoria, B.C., Canada, 1999.

12. Ibid.

13. Allan William Johnson, Jr., "Magic Squares," *Journal of Recreational Mathematics* 19(3):213–216 (1987).

14. John Hendricks, personal communication, and "Hruska's Conjecture" (pamphlet), 1999.

Chapter 2. Classification

1. Martin Gardner, *Time Travel and Other Mathematical Bewilderments*, 218.

2. Mutsumi Suzuki, Japan, http://www.pse.che.tohoku.ac.jp/~msuzuki/MagicSquare.total.html.

3. Martin Gardner, *Time Travel and Other Mathematical Bewilderments*, 218.

4. Martin Gardner, "Magic Squares Cornered," *Nature* 395(6699): 216–217, Sept. 17, 1998.

5. Ibid.

6. Dame Kathleen Ollerenshaw and David Brée, *Most-Perfect Pandiagonal Magic Squares: Their Construction and Enumeration,* Institute of Mathematics and Its Applications, Catherine Richards House, 16 Nelson Street, Southend-on-Sea, Essex SS1 1EF, UK. Also see http: //www.magic-squares.com/.

7. Martin Gardner, "Magic Squares Cornered."

8. The Institute of Mathematics and Its Applications, "Most-Perfect Pandiagonal Magic Squares," http://www.magic-squares.com/pages/quotes/quotes.htm; also see Ian Stewart, "Most-Perfect Magic Squares," *Scientific American* 281(5):122–123 (November 1999).

9. John Lee Fults, *Magic Squares*, 44.

10. W. S. Andrews, *Magic Squares and Cubes*, 209.

11. Emanuel Emanouilidis, "More Magic Squares," *Journal of Recreational Mathematics* 27(3):179–180 (1995).

12. Ibid.

13. W. S. Andrews (*Magic Squares and Cubes*, page 175) says this square was made by Mr. Beverly and published in the *Philosophical Magazine* in 1848.

14. Clifford Pickover, *Mazes for the Mind: Computers and the Unexpected* (New York: St. Martin's Press, 1995), 204.

15. Ibid.

16. John R. Hendricks, "The Magic Hexagram," *Journal of Recreational Mathematics* 25:1 (1993).

17. D. M. Zhou, "A Syllabus of Mathematics and Hui Yang's Methodology of Teaching Mathematics" (Chinese), *Journal of the Central China*

Normal University of Natural Science, 24(3):396–399 (1990); G. Abe, "Magic Squares That Occur in Yang-Hui's Mathematics" (Japanese), *Sugakushi Kenkyu* 70:11–32 (1976). See also: http://www-groups.dcs.st-and.ac.uk/~history/Mathematicians/Yang.html.

18. Martin Gardner, *Time Travel and Other Mathematical Bewilderments*, 219.

19. W. S. Andrews, *Magic Squares and Cubes*.

20. Richard Schroeppel, "Artificial Intelligence Memo 239," MIT, 1972; reported in Martin Gardner's *Time Travel and Other Mathematical Bewilderments*, 219–220.

21. Martin Gardner, "Magic Squares and Cubes (Chapter 17)," in *Time Travel and Other Mathematical Bewilderments,* 213–225.

22. H. Langman, *Play Mathematics* (New York: Hafner, 1962), 75–76.

23. Martin Gardner, *Time Travel and Other Mathematical Bewilderments*, 216–217. John Hendricks, in a personal communication, suggested that the supposed order-7 perfect magic cube has repetitions along one of its diagonals or triagonals.

24. Joseph S. Madachy, *Madachy's Mathematical Recreations*, 101.

25. Ibid., 103.

26. Martin Gardner, *The Second Scientific American Book of Mathematical Puzzles and Diversions* (Chicago: University of Chicago Press, 1987), 137.

27. Ibid., 138.

28. John Hendricks, *Magic Squares to Tesseracts by Computer* (self-published, 1998) 112–113.

29. Joseph Arkin, David C. Arney, and Bruce J. Porter, "The Cameron Cube," *Journal of Recreational Mathematics* 21(2):81–88 (1989).

30. John Hendricks, *Magic Squares to Tesseracts by Computer*, 126.

31. Ibid.

32. W. S. Andrews, *Magic Squares and Cubes*, 241.

33. Eric Weisstein, *CRC Concise Encyclopedia of Mathematics*, 2.

34. Y. I. Perelman, *Fun with Math and Physics* (Moscow: Mir Publishers, 1988).

35. Eric Weisstein, *CRC Concise Encyclopedia of Mathematics*.

36. "Mark Longridge's Rubik's Cube Web Site," http://web.idirect.com/~cubeman/; "Mark Jeay's Cube Page," http://qlink.queensu.ca/~4mj2/rubiks.html.

37. "Mark Jeay's Cube Page," http://qlink.queensu.ca/~4mj2/rubiks.html.

38. Ibid.

39. Eric Weisstein, *CRC Concise Encyclopedia of Mathematics*; Dan Hoey, "The Real Size of Cube Space," http://www.math.rwth-aachen.de/~Martin. Schoenert/Cube-Lovers/Dan_Hoey__The_real_size_of_cube_space.html; Douglas Hofstadter, "Metamagical Themas: The Magic Cube's Cubies Are Twiddled by Cubists and Solved by Cubemeisters," *Scientific American* 244:20–39 (March 1981); M. Larson, "Rubik's Revenge: The Group Theoretical Solution," *American Mathematics Monthly* 92:381–390 (1985); D. Miller, "Solving Rubik's Cube Using the 'Bestfast' Search Algorithm and 'Profile' Tables," http://www.sunyit.edu/~millerd1/RUBIK.HTM; M. Schoenert, "Cube Lovers: Index by Date," http://www.math.rwth-aachen.de/~Martin.Schoenert/Cube-Lovers/; M. Schubart, "Rubik's Cube Resource List," http://www.best.com/~schubart/rc/resources.html; D. Singmaster, "Notes on Rubik's 'Magic Cube,'" (Hillside, N.J.: Enslow, 1981); D. Taylor, *Mastering Rubik's Cube* (New York: Holt, Rinehart and Winston, 1981); D. Taylor and L. Rylands, *Cube Games: 92 Puzzles & Solutions* (New York: Holt, Rinehart and Winston, 1981); E. Turner and K. Gold, "Rubik's Groups," *American Mathematics Monthly* 92:617–629 (1985).

40. Dan Velleman, "Rubik's Tesseract," *Mathematics Magazine* 65(1):27–36, (February 1992).

41. Eric Weisstein, *CRC Concise Encyclopedia of Mathematics*, 16. Also see J. Hunter and J. Madachy, "Mystic Arrays," Chapter 3, in *Mathematical Diversions* (New York: Dover, 1975), 30–31; Joseph S. Madachy, *Madachy's Mathematical Recreations*, 89–91.

42. Eric Weisstein, *CRC Concise Encyclopedia of Mathematics*, 35. Also see L. Saloons, "Alpha Magic Squares," in *The Lighter Side of Mathematics*, edited by R. Guy and R. Woodrow (Washington, D.C.: Mathematics Association of America, 1994). L. Saloons, "Alphamagic Squares," *Abacus* 4, 28–45 (1986); L. Saloons, "Alphamagic Squares 2," *Abacus* 4:20–29, 43 (1987).

43. Ivars Peterson, *Islands of Truth* (Freeman: New York, 1990), 162.

44. Eric Weisstein, *CRC Concise Encyclopedia of Mathematics*. W. W. Ball and H.S.M. Coxeter, *Mathematical Recreations and Essays* (New York: Dover, 1987), 212; J. Hunter and J. Madachy, "Mystic Arrays," 31; M. Kraitchik, "Multimagic Squares," *Mathematical Recreations* (New York: W. W. Norton, 1942) 176–178.

45. Ibid.

46. Ibid.

47. Eric Weisstein, *CRC Concise Encyclopedia of Mathematics*.

48. Ibid.; H. Stapleton, "The Gnomon as a Possible Link Between (a) One Type of Mesopotamian Ziggurat and (b) the Magic Square Numbers on which Jaribian Alchemy Was Based," *Ambix: Journal of the Society for the Study of Alchemy and Early Chemistry* 6:1–9 (1957–1958).

49. Charles Trigg, "A Remarkable Group of Gnomon-Antimagic Squares," *Journal of Recreational Mathematics* 19(3):164–166 (1987).

50. Eric Weisstein, *CRC Concise Encyclopedia of Mathematics*; R. Guy, "Unsolved Problems Come of Age," *American Mathematics Monthly* 96:903–909 (1989). Readers may be interested in the Graceful Tree Conjecture. Let T be a tree (a finite, simple, connected, acyclic graph). We place the integers 1 through n on the vertices of T. On each edge we write the (absolute value of the) difference between the labels on its end points. The labeling is *graceful* if the edge labels have the integers 1 through $n - 1$. It is conjectured that all trees have such a labeling.

51. Martin Gardner, *The Sixth Book of Mathematical Games from Scientific American* (Chicago: University of Chicago Press, 1984), 24.

52. Eric Weisstein, *CRC Concise Encyclopedia of Mathematics*; K. Abraham, *Philadelphia Evening Bulletin*, July 19, 1963, p. 18, and July 30, 1963; M. Beeler, Item 49 in M. Beeler, R. Gosper, R. Schroeppel, and R. Haemem, "Memo AIM-239," (Cambridge, Mass.: MIT Artificial Intelligence Laboratory, February 1972), 18; Martin Gardner, "Permutations and Paradoxes in Combinatorial Mathematics," *Scientific American* 209:112–119 (Aug. 1963); Martin Gardner, *The Sixth Book of Mathematical Games from Scientific American*, 22–24; R. Honsberger, *Mathematical Gems I* (Washington, D.C.: Mathematics Association of America, 1973), 69–76; Joseph S. Madachy, *Madachy's Mathematical Recreations*, 100–101; Charles Trigg, "A Unique Magic Hexagon," *Journal of Recreational Mathematics*, January, 1964; T. Vickers, *Mathematical Gazette*, p. 291, 1958.

53. Eric Weisstein, *CRC Concise Encyclopedia of Mathematics*; J. Hunter and Joseph S. Madachy, "Mystic Arrays," 30–31; Joseph S. Madachy, *Madachy's Mathematical Recreations*, 89–91.

54. Maurice Kraitchik, "Magic Series." §7.13.3 in *Mathematical Recreations* (New York: W. W. Norton, 1942), 183–186. On page 146, Kraitchik defines the nth magic constant to be $n(n^2 + 1)/2$. He writes, "any set of n distinct numbers from 1 to n^2 whose sum is the nth magic constant is called a magic series." On page 176, he generalizes that the nth magic constant of degree p is $1/n$ times the sum of the first n^2 pth powers. For example, the third magic constant of degree 4 would be $(1^4 + 2^4 + \cdots + 9^4)/3$. Then he writes, "$n$ numbers form a magic series of degree p if the sum of their kth powers is the [nth] magic constant of degree k, for every k from 1 to p." Presumably, the n numbers have to

be distinct and between 1 and n^2. Kraitchik doesn't give any fully worked examples, although he gives a construction that should lead to some. According to mathematician Gerry Myerson, $(2, 8, 9, 15)$ is magic, bimagic, and trimagic, which may mean that it is a magic series of degree 3. In a similar analysis of the meaning of "magic series," mathematician Kurt Foster reminds us again that the nth magic constant (of degree 1) is $1/n$ times the sum of the first n^2 positive integers; its value is $n(n^2 + 1)/2$. A magic series is a series n of the integers from 1 to n^2, which add up to the magic constant. Recall that a magic square is an $n \times n$ square array filled with the first n^2 positive integers, each of whose rows, columns, and diagonals have the same sum. This sum must, of course, be the magic constant. So, each row, column, and diagonal of a magic square is a magic series. Some magic squares remain magic (all rows, columns, and diagonals have the same sum) if the entries are replaced by their kth powers. If this occurs, the common sum must be $1/n$ times the sum of the first n^2 kth powers, and this is the nth magic constant of degree k. Such a square is called multimagic.

55. Nora Hartsfield and Gerhard Ringel, "Supermagic and Antimagic Graphs," *Journal of Recreational Mathematics* 21(2):107–115 (1989).

56. Ivan Moscovich, *Fiendishly Difficult Math Puzzles* (New York: Sterling, 1986), 18.

57. Boris Kordemsky, *The Moscow Puzzles* (New York: Dover, 1972), 144.

58. Ian Stewart, "Knotted Ventured," *Scientific American* 283(1):105 (July 2000).

59. Thomas Hagedorn, "On the Existence of n-Dimensional Magic Rectangles," *Discrete Mathematics*, 207(1–3):53–63 (September 1999). Also see Thomas Hagedorn, "Magic Rectangles Revisited," 207(1–3):65–72 (September 1999). Hagedorn clarifies in a personal communication to me, "One can have an 'even' n-dimensional magic rectangle, but it cannot be the case that two of the sides have dimension 2. For example, a magic rectangle of size $(2, 2, 4)$ cannot exist. Nor can those of size $(2, 2, x, y, w, z, \ldots)$." The theorems in his papers include: (1) for m, $n > 1$, there is a $m \times n$ rectangle if and only if $m \equiv n$ mod 2 and $(m, n) \neq (2, 2)$; (2) If m_i are positive even integers with $(m_i, m_j) \neq (2, 2)$ for $i \neq j$, then a magic n-rectangle of size $(m_i, \ldots, m_n)$ exists. Note that magic rectangles were first extensively studied as far back as 1881.

Chapter 3. Gallery I: Squares, Cubes, and Tesseracts

1. W. S. Andrews, *Magic Squares and Cubes*, 89. In 2001, Paul Pasles, a number theorist at Villanova University in Philadelphia, discovered several more Franklin squares with four, six, eight, and sixteen numbers to a

side. The new findings suggest that Franklin used at least four different methods for constructing the squares. It is amazing that these new squares, discovered in a letter Franklin wrote in 1765 and from other sources, did not come to light until the twenty-first century. For more information, see Constance Holden, "Number Fun with Ben," *Science* 292(5518):843 (May 4, 2001).

2. Christopher J. Henrich, "Magic Squares and Linear Algebra," *American Mathematics Monthly*, 98(6):481–488 (1991). See especially page 486.

3. For those of you not familiar with numbers represented in bases other than 10 (which is the standard way of representing numbers), consider how to represent any number in base 2. Numbers in base 2 are called binary numbers. To represent a binary number, only the digits 0 and 1 are used. Each digit of a binary number represents a power of 2. The rightmost digit is the 1s digit, the next digit to the left is the 2s digit, and so on. In other words, the presence of a 1 in a digit position indicates that a corresponding power of 2 is used to determine the value of the binary number. A 0 in the number indicates that a corresponding power of 2 is absent from the binary number. An example should help. The binary number 1111 represents $(1 \times 2^3) + (1 \times 2^2) + (1 \times 2^1) + (1 \times 2^0) = 15$. The binary number 1000 represents $1 \times 2^3 = 8$. Here are the first eight numbers represented in binary notation: 0000, 0001, 0010, 0011, 0100, 0101, 0110, and 0111. It turns out that any number can be written in the form $c_n b^n + c_{n-1} b^{n-1} + \cdots + c_2 b^2 + c_1 b^1 + c_0 b^0$, where b is a base of computation and c is some nonnegative integer less than the base.

4. Jim Moran, *The Wonders of Magic Squares* (New York: Vintage Books, 1982).

5. Lalbhai Patel, "The Secret of Franklin's 8×8 'Magic' Square," *The Journal of Recreational Mathematics* 23(3):175–182 (1991).

6. W. S. Andrews, *Magic Squares and Cubes*.

7. Joseph S. Madachy, *Madachy's Mathematical Recreations*, 94.

8. Harvey Heinz, "Unusual Magic Squares of Patrick De Geest," http://www.geocities.com/CapeCanaveral/Launchpad/4057/ unususqr.htm#U; see also Patrick De Geest, "World of Numbers," http://ping4.ping.be/~ping6758//index.shtml.

9. Rodolfo Kurchan, "An All Pandigital Magic Square (question posed by Rudolf Ondrejka)," *Journal of Recreational Mathematics* 23(1):69–78 (1991). Rodolfo Kurchan, "Rodolfo Kurchan's web page," http:// www.geocities.com/TimesSquare/Maze/1320/rodolfo/rodolfo.html#inicio. In his own words: "*Mi nombre es Rodolfo Marcelo Kurchan y nací en Buenos Aires, Argentina el 21 de marzo de 1971. Mi primer contacto con los acertijos es a*

los 12 años cuando mis padres me regalan el libro de Martin Gardner, "Para-dojas Ajá."

10. S. Boardman, "Theta Problem Page," *Theta* 10(2):25 (1996).

11. Clifford Pickover, *Surfing Through Hyperspace* (New York: Oxford University Press, 2000).

12. John Hendricks, "The Magic Tesseracts of Order 3 Complete," *Journal of Recreational Mathematics* 22(1):16–26 (1990); John Hendricks, "The Five- and Six-Dimensional Magic Hypercubes of Order 3," *Canadian Mathematical Bulletin* 5(2):171–189 (May 1962); John Hendricks, "Magic Tesseract," in *The Pattern Book: Fractals, Art, and Nature*, Clifford Pickover, ed. (River Edge, N. J.: World Scientific, 1995).

13. J. Denes and A. Keedwell, *Latin Squares and Their Applications* (New York: Academic Press, 1974); J. Denes and A. Keedwell, *Latin Squares: New Developments in the Theory and Applications* (Amsterdam: North-Holland, 1991). Also see Eric Weisstein, "Latin Square," http://mathworld.wolfram.com/ LatinSquare.html. Other Latin square references: R. Bose and B. Manvel, *Introduction to Combinatorial Theory* (Ch. 7) (New York: John Wiley & Sons, 1984), 135–149; M. Jacobson and P. Matthews, "Generating Uniformly Distributed Latin Squares," *Journal of Combinatorial Designs* 4(6):405–437 (1996); B. McKay and E. Rogoyski, "Latin Squares of Order 10," *Electronic Journal of Combinatorics* 2(3):1–4 (1995); J. Shao and W. Wei, "A Formula for the Number of Latin Squares," *Discrete Mathematics* 110:293–296 (1992); J. Byers, "Basic Algorithms for Random Sampling and Treatment Randomization," *Computers in Biology and Medicine* 21(112):69–77 (1991); Charles F. Laywine and Gary L. Mullen, *Discrete Mathematics Using Latin Squares* (New York: John Wiley & Sons, 1998)(includes Latin hypercubes and practical examples).

Several excellent papers focus on *transversals* of Latin squares: Sherman Stein, "Transversals of Latin Squares and their Generalizations," *Pacific Journal of Mathematics* 59(2):567–575 (1975); Paul Erdös, D. R. Hickerson, D. A. Norton, and Sherman Stein, "Has Every Latin Square of Order n a Partial Latin Transversal of Size $n-1$?" *American Mathematics Monthly* 95(5):428–430 (May 1988); and P. W. Shor, "A Lower Bound for the Length of a Partial Transversal of a Latin Square," *Journal of Combinatorial Theory Series A* 33:1–8 (1982). Stein has introduced several types of arrays, including the equi-n-square (an n by n array in which each symbol occurs exactly n times), and established the existence of transversals with many distinct elements. For example, the equi-n-square has a transversal with at least

$$n\left(1 - \frac{1}{2!} + \frac{1}{3!} - \cdots \pm \frac{1}{n!}\right) \approx (1 - 1/e)n \approx 0.63n$$

distinct symbols. P. W. Shor showed that every Latin square has a partial Latin transversal of length at least $n - 5.53(\ln n)^2$. Transversal guru Sherman Stein recently wrote to me, "I conjecture that in any $n - 1$ by n array in which each symbol occurs at most $n - 1$ times there is a Latin transversal of $n - 1$ entries (no symbol, row, or column duplicated). Perhaps an amateur with a computer could find a counterexample."

14. Martin Gardner, *Fractal Music, Hypercards, and More . . .* (New York: Freeman, 1992).

15. John Lee Fults, *Magic Squares*, 93.

16. Harvey Heinz, http://www.geocities.com/CapeCanaveral/ Launchpad/4057/moremsqrs.htm#M.

17. Ibid.

18. Martin Gardner, *Time Travel and Other Mathematical Bewilderments*, 221.

19. Ibid.

20. Ibid., 223.

21. Ibid.

22. Ibid., 224.

23. Ibid.

24. John Hendricks, *The Magic Square Course* (self-published), 419–431; see also "Harvey Heinz's web page," http://www.geocities.com/ CapeCanaveral/Launchpad/4057/hendricks.htm.

25. "F. Poyo's web page," http://makoto.mattolab.kanazawa-it.ac.jp/ ~poyo/index.html.

26. Ibid.

27. Joseph S. Madachy, *Madachy's Mathematical Recreations*, 89.

28. Ibid., 90.

29. Ibid., 92.

30. Ibid., 93.

31. Bureau of Justice Statistics Correctional Surveys, *The National Probation Data Survey, National Prisoner Statistics, Survey of Jails,* and *The National Parole Data Survey.*

32. Ibid., 95; see also H. E. Dudeney, *Amusements in Mathematics* (1917; reprint, New York: Dover, 1970), 125.

33. Eric Weisstein, *CRC Concise Encyclopedia of Mathematics*, 1129.

34. Allan Johnson, "Minimum Prime Pandiagonal Order-6 Magic Squares," *Journal of Recreational Mathematics* 23(3):190–101 (1991).

35. Joseph S. Madachy, *Madachy's Mathematical Recreations,* 95; see also H. E. Dudeney, *Amusements in Mathematics,* 125.

36. W. S. Andrews, *Magic Squares and Cubes,* 173.

37. Ibid., 282.

38. Joseph S. Madachy, *Madachy's Mathematical Recreations,* 109.

39. Harvey Heinz, http://www.geocities.com/CapeCanaveral/ Launchpad/4057/magicsquare.htm.

40. Joseph S. Madachy, *Madachy's Mathematical Recreations,* 110.

41. Ibid., 113.

42. W. S. Andrews, *Magic Squares and Cubes,* 98.

43. Ibid., 28.

44. Ibid., 28.

45. Paul Carus, in W. S. Andrews, *Magic Squares and Cubes,* 115.

46. Paul Carus, *The Gospel of Buddha.*

47. Paul Carus, in W. S. Andrews, *Magic Squares and Cubes,* 118.

48. Ernst F. F. Chladni, *Entdeckungen über die Theorie des Klanges* [Discoveries concerning the theory of sound] (Leipzig, 1787); also see the web sites

   http://mobydick.physics.utoronto.ca/chladni.html

   http://www.sil.si.edu/exhibits/artistsbook/94-13490.jpg

   http://www.alphaomega.se/english/cymatics.html

49. Cathie E. Guzzetta, "Music Therapy: Nursing the Music of the Soul," in *Music: Physician for the Times to Come,* Don Campbell, ed. (New York: Quest Books, 1991), 149.

50. Ian Stewart, *Another Fine Math You've Got Me Into . . .* (New York: Freeman, 1992), 95.

51. Ibid., 96.

52. Ibid., 97.

53. Ibid., 96–97.

54. Ibid., 109.

55. Henry Dudeney, *Amusements in Mathematics.* The general method of making a tour on a board of order $4k + 1$ comes from Maurice Kraitchik, *Mathematical Recreations,* 264–265.

56. Ian Stewart, *Another Fine Math You've Got Me Into . . . ,* 105–106.

57. Edward Falkener, *Games, Ancient and Oriental* (New York and London: Longmans Green & Co., 1982), 300; see also W. S. Andrews, *Magic Squares and Cubes,* 164.

58. Edward Falkener, *Games, Ancient and Oriental*, 337; see also W. S. Andrews, *Magic Squares and Cubes*, 164.

59. W. S. Andrews, *Magic Squares and Cubes*, 167.

60. Ibid., 169.

61. Ibid., 170.

62. Ibid., 171.

63. Ibid., 172.

64. Ibid. In 1993, John Hendricks published the solution to Frierson's fuddle. To correct Frierson's flawed square, interchange the horizontal pairs 65-17 and 64-18. Interchange the vertical pairs 33-49 with 68-14, and 4-78 with 37-45. Perhaps this is the square that Frierson intended. The correction is published in John Hendricks, "Frierson Fuddle," *Journal of Recreational Mathematics* 25(1):77 (1993).

65. Ibid., 173.

66. Joseph S. Madachy, *Madachy's Mathematical Recreations*, 88.

67. W. W. Ball and H. S. M. Coxeter, *Mathematical Recreations and Essays*, 185–187.

68. Ed Pegg Jr., http://www.mathpuzzle.com/leapers.htm; Donald Knuth, http://www-cs-faculty.stanford.edu/~knuth/preprints.html.

69. Stanely Rabinowitz, "A Magic Rook's Tour," *Journal of Recreational Mathematics* 18(3):203–104 (1985–86).

70. W. S. Andrews, *Magic Squares and Cubes*, 177.

71. Ibid., 174–187.

72. Ibid., 213.

73. Harvey Heinz, http://www.geocities.com/CapeCanaveral/Launchpad/4057/magicsquare.htm; see also John R. Hendricks, *Magic Square Course*, 290–294.

74. Ibid.

75. Personal communication with John Hendricks, Victoria, B.C. Canada.

76. John Hendricks, *Inlaid Magic Squares and Cubes* (self-published, 1999), 20.

77. Ibid., 34.

78. Ibid., 54.

79. Ibid., 77.

80. Ibid., 102.

81. Ibid., A8.

82. Harvey Heinz, http://www.geocities.com/CapeCanaveral/Launchpad/4057/hendricks.htm (Square by John Hendricks). Also see John R. Hendricks, *Magic Square Course*, 244.

83. W. S. Andrews, *Magic Squares and Cubes*, 214.

84. Ibid., 240.

85. W. W. Ball and H. S. M. Coxeter, *Mathematical Recreations and Essays*, 212.

86. The 1750 date for this square is probably apocryphal, and I suspect a more recent origin.

87. See the web sites

http://www.geocities.com/CapeCanaveral/Launchpad/4057/MoreMsqrs.htm

http://www.grogono.com/magic/

http://grogono.com/magic/9 × 9.shtml

88. E. W. Shineman, Jr., "The 41 Magic Square," *Recreational & Educational Computing* (newsletter), 8(1/2):5 (July 1993). (More information on this newsletter is at http://members.aol.com/DrMWEcker/REC.html; see also http://www.geocities.com/CapeCanaveral/Launchpad/4057/REC.htm#Order-5.)

89. Carlos B. Rivera's "Prime Puzzles Page," http://www.sci.net.mx/~crivera/; see also http://www.geocities.com/CapeCanaveral/Launchpad/4057/UnusuSqr.htm.

90. Harvey Heinz, http://www.geocities.com/CapeCanaveral/Launchpad/4057/unususqr.htm.

91. These patterns were left on my desk by some unknown colleague. Their origin is a mystery. I would be happy to hear from any reader who may find a literature reference for these patterns.

92. Harm Derksen's web page, http://www.math.unibas.ch/~hderksen/magic.html.

93. Dave Harper's web page, http://web.idirect.com/~recmath/ and http://web.idirect.com/~recmath/fset7.html.

94. The Rekord-Klub Saxonia was founded in 1988. There is a strict rule that every member must be a world-record breaker. At the moment, the club members have established 112 world records such as creating the largest bicycle or the longest noodle, or participating in a computer game marathon. Most of their members are mentioned in the *Guinness Book of World Records*. For more information, see Ralf Laue, "Rekord-Klub Saxonia," http://www.imn.htwk-leipzig.de/~saxonia/records/magic.html.

95. John Hendricks, personal communication, and *The Magic Square Course*, 433–456.

96. Ibid., 456.

97. Ibid.

98. Allan William Johnson, "Related Magic Squares," *Journal of Recreational Mathematics*, 20(1):26–27 (1988).

99. Onur Demirors, Nader Rafraf, and Murat Tanik, "Obtaining *N*-Queens Solutions from Magic Squares and Constructing Magic Squares from *N*-Queens Solutions," *Journal of Recreational Mathematics* 24(4):272–280 (1992).

100. Ibid.

101. Ibid.

102. Allan Johnson, Jr., "Palindromes and Magic Squares," *Journal of Recreational Mathematics* 21(2):97–100 (1989).

103. Henry Dudeney, *536 Curious Problems and Puzzles* (New York: Barnes and Nobles Books, 1995), 142.

104. Ibid., 143.

105. Gary Adamson, San Diego, California, personal communication, April 1999. Other authors have discussed binary codes and DNA. For example, see Johnson F. Yan, *DNA and the I-Ching* (New York: North Atlantic Books, 1991); Katya Walter, *Tao of Chaos: Merging East and West* (New York: Kairos Center, 1994); Clifford Pickover, "DNA Vectorgrams: Representation of Cancer Gene Sequences as Movements along a 2-D Cellular Lattice," *IBM Journal of Research and Development* 31:111–119 (1987); Clifford Pickover, "Frequency Representations of DNA Sequences: Application to a Bladder Cancer Gene," *Journal of Molecular Graphics* 2:50 (1984); Clifford Pickover, *Computers, Pattern, Chaos and Beauty* (New York: St. Martin's Press, 1990).

106. The pattern was described by a "Mr. Devedec" in the early 1900s. In order to create a magic square of any even order, we start with the numbers written in what will be called the *fundamental position*, that is to say, written in a square, in order, from left to right and up. This means the bottom row of the square contains the numbers $1, 2, 3, \ldots, n$; the row above contains the numbers $n+1, n+2, \ldots, 2n$, etc. The numbers in the magic square represented in Figure 3.61 are replaced by symbols, where O is an element that is in the same position as in the fundamental square; \ is an element that, in the fundamental square, occupied the position symmetric to this with respect to the center; | is an element that, in the fundamental square, occupied the position symmetric to this with respect to the horizontal median; and – indicates an element that, in the

fundamental square, occupied the position symmetric to this with respect to the vertical median. The center is to be filled with either of the bottom two L-shaped pieces depending on whether the square is of order $4k$ or $4k + 2$. For more information, see Maurice Kraitchik, *Mathematical Recreations,* 150–153.

107. Michael Keith, personal communication.

108. Jeremiah Farrell, "Magic Dice," in *The Mathemagician and Pied Puzzler: A Collection in Tribute to Martin Gardner,* Elwyn Berlekamp and Tom Rodgers, eds. (New York: A. K. Peters, 1999).

Chapter 4. Gallery 2: Circles and Spheres

1. Joseph S. Madachy, *Madachy's Mathematical Recreations,* 97.

2. W. S. Andrews, *Magic Squares and Cubes,* 322.

3. Ibid., 323.

4. Ibid., 324.

5. Ibid., 326.

6. Ibid., 326.

7. Ibid., 328.

8. Ibid., 329.

9. Abu 'Uthman' Amr ibn Bakr al-Kinani al-Fuqaimi al-Basri al-Jahiz was born in Basra, Iraq, in A.D. 776. Al-Jahiz was a fascinating man, writing more than two hundred works on topics ranging from zoology to Arabic grammar, poetry, rhetoric, and lexicography. Some of his famous books are *The Book of Animals, The Art of Keeping One's Mouth Shut, Against Civil Servants, Arab Food, In Praise of Merchants,* and *Levity and Seriousness.* His most famous book, *Kitab al-Hayawan (Book of Animals),* was an encyclopedia of seven large volumes. He was rewarded with 5000 gold dinars from the court official to whom he dedicated the *Book of Animals.* Al-Jahiz died in Basra in 868 as a result of an accident in which he was crushed to death by a collapsing pile of books in his private library. (For more information see http://www.erols.com/gmqm/jahiz.html.)

10. Harvey Heinz, http://www.geocities.com/CapeCanaveral/Launchpad/4057/moremsqrs.htm.

11. W. S. Andrews, *Magic Squares and Cubes,* 332.

12. Ibid., 333.

13. Ibid., 335.

14. Ibid., 336.

15. Ibid., 334–336.

16. Ibid., 337.

17. Ibid., 330.

18. Clifford Pickover, http://sprott.physics.wisc.edu/pickover/home.htm.

19. Michael Keith, personal communication.

20. Ibid.

21. Ibid.

22. Ibid.

23. Clifford Pickover, http://www.pickover.com.

24. Ibid.

25. Ibid.

Chapter 5. Gallery 3: Stars, Hexagons, and Other Beauties

1. Joseph S. Madachy, *Madachy's Mathematical Recreations,* 99.

2. W. S. Andrews, *Magic Squares and Cubes,* 341.

3. Ibid., 342–350.

4. Joseph S. Madachy, *Madachy's Mathematical Recreations,* 101.

5. Ibid., 112.

6. W. W. Ball and H.S.M. Coxeter, *Mathematical Recreations and Essays,* 213–214.

7. Ibid., 214.

8. W. Ball and H.S.M. Coxeter, *Mathematical Recreations and Essays,* 216.

9. Ibid.

10. Harvey Heinz, http://www.geocities.com/CapeCanaveral/Launchpad/4057/moremsqrs.htm.

11. Ibid. See also Martin Gardner, *Sixth Book of Mathematical Games from Scientific American* (Chicago: Chicago University Press, 1970), 22–23.

12. William Benson and Oswald Jacoby, *New Recreations with Magic Squares* (New York: Dover, 1976), also see Harvey Heinz, http://www.geocities.com/CapeCanaveral/Launchpad/4057/UnusuSqr.htm#U.

13. Clifford Pickover, www.pickover.com.

14. Leisa Holiday, http://www.beachworks.com/intuitive/durga.html.

15. PennyLea Morris Seferovich, http://www.americansanskrit.com/inspire/yantras.html.

16. Clifford Pickover, www.pickover.com.

17. Durga represented a vision of the feminine that challenged the submissive female stereotype in traditional societies and religions. Her turbulent story was one of war, success, personal sacrifice, and final liberation. Mother goddess in Hindu mythology, she rides a tiger and carries the weapons of all the gods in her many arms. She is praised as being responsible for the creation, sustenance, and withdrawal of the universe. She combines beauty and terror, a mysterious power, the personification of knowledge, wisdom, and memory.

Even today, goddess images are among the most divine and powerful in the Hindu iconography. The great Devi (Mother) as the genetrix of all things is not only seen as a benign nurturer of life, but she can be wildly unpredictable, bringing disease, flood, famine, and death. Her awesome power is represented especially in her manifestation as Kali, "the Black One," with her angry red lolling tongue. In her many spinning arms, she carries ritual weapons and a human skull for catching sacrificial blood.

In classical Hindu mythology the goddess Durga is one of the commonest names of Lord Shiva's consort. Other names are Uma, Devi, Gauri, Parvati, Chandi, Chamunda, Kali, Shakti, Bhavani, Ambika, etc. She is revered for her victories over several Asures (demons), primarily the buffalo demons Mahishasur, Shumbha, Nishumbha, Chandh, and Mundh.

18. Clifford Pickover, www.pickover.com.

19. John Hendricks, "A Note on Magic Tetrahedrons," *Journal of Recreational Mathematics* 24(4):245–249 (1992).

20. Boris Kordemsky, *The Moscow Puzzles* (New York: Dover, 1972), 144.

21. Ibid., 144.

22. Ibid., 146.

23. Ibid., 146.

24. Clifford Pickover, www.pickover.com.

25. Harvey Heinz, http://www.geocities.com/CapeCanaveral/Launchpad/4057/unususqr.htm.

26. Michael Keith, personal communication.

27. Harvey Heinz, http://www.geocities.com/CapeCanaveral/Launchpad/4057/primestars.htm.

28. Clifford Pickover, www.pickover.com.

29. Nagle Designs by Nancy Nagle, http://www.nagledesign.com/.

30. James Nesi (a.k.a. Twisto), "Fubine," *The Enigma* 1128 (March 1998). (*The Enigma* is a monthly magazine of the National Puzzler's League (http://www.puzzlers.org), a nonprofit education organization founded in 1883.)

Some Final Thoughts

1. Martin Gardner, *Order and Surprise* (Amherst, N.Y.: Prometheus Books, 1983).

2. I'm sure that most mathematicians do not look down on recreational mathematics or those who enjoy it. Recreational mathematics is what "hooked" many mathematicians into the subject.

For Further Reading

W. S. Andrews, *Magic Squares and Cubes* (1917; reprint, New York: Dover, 1960).

W. W. Ball and H.S.M. Coxeter, *Mathematical Recreations and Essays* (New York: Dover, 1987).

H. E. Du, "Magic Squares," *Encyclopedia Britannica* (Chicago: Encyclopedia Britannica, 1962), 14: 62.

John Lee Fults, *Magic Squares* (La Salle, Illinois: Open Court, 1974).

Martin Gardner, *Time Travel and other Mathematical Bewilderments* (New York: Freeman, 1987)

Jan Gullberg, *Mathematics: From the Birth of Numbers* (New York: W.W. Norton, 1997).

John Hendricks, *Inlaid Magic Squares and Cubes* (self-published, 1999). (John Hendricks, the guru of magic squares and cubes, has published a series of wonderful books, including *Magic Squares to Tesseracts by Computer* and *Third-Order Magic Tesseracts*. You can write to him for more information at John Hendricks, #308-151 St. Andrews Street, Victoria, B.C., V8V 2M9, Canada.)

Christopher J. Henrich, *"Magic Squares and Linear Algebra,"* American Mathematics Monthly, 98(6): 481–488(1991).

Annemarie Schimmel, *The Mystery of Numbers* (Oxford University Press: New York, 1993).

Eric Weisstein, *CRC Concise Encyclopedia of Mathematics* (New York: CRC Press, 1998).

Index

About the Author

Clifford A. Pickover received his Ph.D. from Yale University's Department of Molecular Biophysics and Biochemistry. He graduated first in his class from Franklin and Marshall College, after completing the four-year undergraduate program in three years. His many books have been translated into Italian, French, Greek, German, Japanese, Chinese, Korean, Portuguese, and Polish. He is author of the popular books *The Paradox of God and the Science of Omniscience* (Palgrave, 2002), *The Stars of Heaven* (Oxford University Press, 2001), *Dreaming the Future* (Prometheus, 2001), *Wonders of Numbers* (Oxford University Press, 2000), *The Girl Who Gave Birth to Rabbits* (Prometheus, 2000), *Surfing Through Hyperspace* (Oxford University Press, 1999), *The Science of Aliens* (Basic Books, 1998), *Time: A Traveler's Guide* (Oxford University Press, 1998), *Strange Brains and Genius: The Secret Lives of Eccentric Scientists and Madmen* (Plenum, 1998), *The Alien IQ Test* (Basic Books, 1997), *The Loom of God* (Plenum, 1997), *Black Holes: A Traveler's Guide* (Wiley, 1996), and *Keys to Infinity* (Wiley, 1995). He is also the author of numerous other highly acclaimed books, including *Chaos in Wonderland: Visual Adventures in a Fractal World* (1994), *Mazes for the Mind: Computers and the Unexpected* (1992), *Computers and the Imagination* (1991), and *Computers, Pattern, Chaos, and Beauty* (1990)—all published by St. Martin's Press—as well as the author of over 200 articles concerning topics in science, art, and mathematics. He is also coauthor, with Piers Anthony, of *Spider Legs,* a science fiction novel once listed as Barnes and Noble's second-best-selling science fiction title.

Pickover is currently an associate editor for the scientific journals *Computers and Graphics* and *Theta Mathematics Journal* and is an editorial board member for *Odyssey, Idealistic Studies, Leonardo,* and *YLEM.* He has been a guest editor for several scientific journals.

Dr. Pickover is editor of the books *Chaos and Fractals: A Computer Graphical Journey* (Elsevier, 1998); *The Pattern Book: Fractals, Art, and Nature* (World Scientific, 1995); *Visions of the Future: Art, Technology, and Computing in the Next Century* (St. Martin's Press, 1993); *Future Health* (St. Martin's Press, 1995); *Fractal Horizons* (St. Martin's Press, 1996); and *Visualizing Biological Information* (World Scientific, 1995). He is coeditor of the books *Spiral Symmetry* (World Scientific, 1992) and *Frontiers in Scientific Visualization* (Wiley, 1994). Dr. Pickover's primary interest is finding new ways to continually expand creativity by melding art, science, mathematics, and other seemingly disparate areas of human endeavor.

The *Los Angeles Times* recently proclaimed, "Pickover has published nearly a book a year in which he stretches the limits of computers, art and thought." Pickover received first prize in the Institute of Physics' "Beauty of Physics Photographic Competition." His computer graphics have been featured on the covers of many popular magazines, and his research has recently received considerable attention by the press—including CNN's "Science and Technology Week," the Discovery Channel, *Science News, The Washington Post, Wired,* and the *Christian Science Monitor*—and also in international exhibitions and museums. *OMNI* magazine recently described him as "Van Leeuwenhoek's twentieth century equivalent." *Scientific American* has featured his graphic work several times, calling it "strange and beautiful, stunningly realistic." *Wired* magazine wrote, "Bucky Fuller thought big, Arthur C. Clarke thinks big, but Cliff Pickover outdoes them both." Among his many patents, Pickover has received U.S. Patents 5,095,302 for a 3-D computer mouse, 5,564,004 for strange computer icons, and 5,682,486 for black-hole transporter interfaces to computers.

Dr. Pickover is currently a Research Staff Member at the IBM T. J. Watson Research Center, where he has received thirty-five invention achievement awards, three research division awards, and six external honor awards. Dr. Pickover has been the lead columnist for the Brain-Boggler column in *Discover* magazine for many years and currently writes the Brain-Strain column for *Odyssey* and for studyworksonline.com. His calendar and card sets, *Mind-Bending Visual Puzzles,* are among his most popular creations.

Dr. Pickover's hobbies include the practice of Ch'ang-Shih Tai-Chi Ch'uan and Shaolin Kung Fu, raising golden and green severums (large Amazonian fish), and piano playing (mostly jazz). He is also a member of the SETI League, a group of signal processing enthusiasts who systematically search the sky for intelligent extraterrestrial life. Visit his web site, which has received over 500,000 visits: http://www.pickover.com. He can be reached at P.O. Box 549, Millwood, New York 10546-0549 USA.

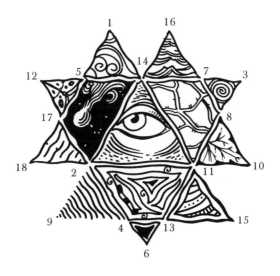